Grade Two S<

Aligned to Alberta Curriculum

Written by Tracy Bellaire and Andrew Gilchrist

This teacher resource has been designed to give students an understanding and appreciation of the 5 topics in Grade 2 Science Alberta Curriculum. Those 5 topics are Exploring Liquids, Buoyancy and Boats, Magnetism, Hot and Cold Temperatures and Small Crawling and Flying Animals. The lessons are designed to involve tactile participation and knowledge application while providing opportunities to connect ideas between topics and school subjects.

TRACY BELLAIRE is an experienced teacher who continues to be involved in various levels of education in her role as Differentiated Learning Resource Teacher in an elementary school in Ontario. She enjoys creating educational materials for all types of learners, and providing tools for teachers to further develop their skill set in the classroom. She hopes that these lessons help all to discover their love of science!

ANDREW GILCHRIST taught primary and junior classes for the Hastings Prince Edward District School Board in Ontario. He earned his teaching degree with a Teaching Subject qualification in History. He recently co-authored a series of mathematics programs that focus on fluency skills with basic math operations and fractions. The series, used for children with special education needs, was used for a study with University of Ottawa. Students using the program demonstrated achievements in fluency at twice the rate of regular in-class instruction, allowing them to catch up with peers.

Copyright © On The Mark Press 2016

Some material appearing in this book has been used in other published works, such as Earth and Space Science Grade 2 (OTM2153), Physical Science Grade 2 (OTM2145) and Magnets (OTM2128).

Published in Canada by:
On The Mark Press
15 Dairy Avenue, Napanee, Ontario, K7R 1M4
www.onthemarkpress.com

Funded by the
Government
of Canada

AT A GLANCE

Skills: Science Inquiry

2.1 Investigate, with guidance, the nature of things, demonstrating an understanding of the procedures followed. Students will identify one or more possible answers, predictions or hypotheses to questions asked by themselves and others.

2.2 Recognize pattern and order in objects and events studied; and, with guidance, record procedures and observations, using pictures and words; and make predictions and generalizations, based on observations. Students will use materials and make observations relevant to the questions asked, carry out simple procedures and identify materials used.

2.3 Construct, with guidance, an object that achieves a given purpose, using materials that are provided. Construction tasks will involve building stable objects that float on water. Students will identify the purpose of the object to be constructed through questions like "What structure do we need to make?" or "What does it accomplish?"

Attitudes

2.4 Demonstrate positive attitudes to the study of science and to the application of science. Students will show growth in acquiring and applying curiosity, inventiveness, perseverance, appreciation of the value of experience and observation, a willingness to work with others, a sense of responsibility for actions taken and respect for living things and environments with a commitment for their care.

Topic A: Exploring Liquids

2.5 Describe some properties of water and other liquids, and recognize the importance of water to living and nonliving things. Students will recognize and describe the characteristics of liquids in terms of the flow of a liquid, the shape of drops and the surface of calm water. Students will recognize that water is a component of many materials and living things and understand human responsibilities for maintaining clean supplies of water while identifying actions taken to ensure safe water supplies.

2.6 Describe the interaction of water with different materials, and apply that knowledge to practical problems of drying, liquid absorption and liquid containment. Students will compare water with other liquids based on colour, ease of flow, tendency of drops to form a ball or bead shape and how the liquids interact with other liquids and solid materials. Students will compare the amount of liquid absorbed by different materials such as paper, sponge, fabric and plastic and evaluate the suitability of different materials for containing liquids by observing what materials can hold water. Students will demonstrate that liquid water can be changed to other states through heating and cooling, predict that a wet surface will dry more quickly when exposed to wind or heating and apply this understanding to practical situations such as drying fabrics or materials and predict the water level in open containers will decrease due to evaporation but the level in close containers will stay the same.

Topic B: Buoyancy and Boats

2.7 Construct objects that will float on and move through water, and evaluate various designs of watercraft. Students will describe, classify and order materials on the basis of buoyancy in order to distinguish between materials that sink in water and those that float; alter or add to a floating object so that it will sink as well as add or alter a non-floating object to make it float; assemble materials with the purpose of having them float, carry a load and be stable in water. Students will develop and adapt methods of construction appropriate to the design task; change the design of a watercraft so it can be propelled through water; evaluate and explain the appropriateness of various materials to the construction of watercraft through considering the degree to which a material is waterproof or can make waterproof joints with other parts, the stiffness or rigidity of the material and the buoyancy of the material.

OTM-2171   ISBN: 978-1-4877-0199-4 © On The Mark Press

Topic C: Magnetism

2.8 Describe the interaction of magnets with other magnets and with common materials. Students will identify where magnets are used in the environment and why they are used; distinguish materials that are attracted by a magnet from those that are not, predict what metallic and non-metallic items will be attracted by a magnet while recognizing that magnets attract materials with iron and steel parts; recognize magnets have polarity and demonstrate that poles repel or attract each other, state a rule for when poles will repel or attract each other, design and produce a device that uses a magnet; demonstrate that some materials will interact with a magnet while other materials will be transparent to effects of a magnet.

Topic D: Hot and Cold Temperature

2.9 Recognize the effects of heating and cooling and identify methods for heating and cooling. Students will describe temperature in relative terms using expressions such as "hotter than" and "colder than", measure temperature in degrees Celsius (C), describe how heating and cooling materials can change them, identify safe practises for handling hot and cold materials in order to avoid potential dangers, recognize the human body is relatively constant while a change in temperature often signals a change in health. Students will identify ways in which the temperature in homes and buildings can be adjusted, describing how local buildings are heated by identifying energy sources, air or water circulation systems, school building heating and cooling systems. Students will describe the role of insulation in keeping things hot or cold, identify places where some form of insulation is used, identify materials that insulate animals from the cold, design and construct a device to keep something hot or cold and describe ways in which temperature changes affect us in our daily lives.

Topic E: Small Crawling and Flying Animals

1.10 Describe the general structure and life habits of small crawling and flying animals and apply this knowledge to interpret local species that have been observed. Students will recognize that there are many kinds of small crawling and flying animals, identify a range of examples found locally, compare and contrast small local animals, including invertebrates, recognize that small animals have homes where they meet their basic needs of air, food, water, shelter and space and describe any special characteristics that help the animals survive. Students will identify each animal's role in the food chain and identify ways in which animals are considered helpful or harmful to humans and the environment, identifying animals as plant eaters, animal eaters or decomposers. Students will describe the relationship of these animals to other living and non-living things in their habitat, give examples of ways that small animals avoid predators including the use of cover, camouflage, flight and keen senses and recognize how some plants and animals must adapt to extreme conditions to meet basic needs. Students will describe conditions for the care of small animals and demonstrate responsible care in maintaining a small animal for a number of days.

Taken from the Alberta Education Grade 2 Science Curriculum.

TABLE OF CONTENTS

OTM-2171   ISBN: 978-1-4877-0199-4 © On The Mark Press

Teacher Assessment Rubric

Student's Name: _____   Date: _____

Success Criteria	Level 1	Level 2	Level 3	Level 4
Knowledge and Understanding Content				
Demonstrate an understanding of the concepts, ideas, terminology definitions, procedures and the safe use of equipment and materials	Demonstrates limited knowledge and understanding of the content	Demonstrates some knowledge and understanding of the content	Demonstrates considerable knowledge and understanding of the content	Demonstrates thorough knowledge and understanding of the content
Thinking Skills and Investigation Process				
Develop hypothesis, formulate questions, select strategies, plan an investigation	Uses planning and critical thinking skills with limited effectiveness	Uses planning and critical thinking skills with some effectiveness	Uses planning and critical thinking skills with considerable effectiveness	Uses planning and critical thinking skills with a high degree of effectiveness
Gather and record data, and make observations, using safety equipment	Uses investigative processing skills with limited effectiveness	Uses investigative processing skills with some effectiveness	Uses investigative processing skills with considerable effectiveness	Uses investigative processing skills with a high degree of effectiveness
Communication				
Organizes and communicates ideas and information in oral, visual, and/or written forms	Organizes and communicates ideas and information with limited effectiveness	Organizes and communicates ideas and information with some effectiveness	Organizes and communicates ideas and information with considerable effectiveness	Organizes and communicates ideas and information with a high degree of effectiveness
Use science and technology vocabulary in the communication of ideas and information	Uses vocabulary and terminology with limited effectiveness	Uses vocabulary and terminology with some effectiveness	Uses vocabulary and terminology with considerable effectiveness	Uses vocabulary and terminology with a high degree of effectiveness
Application of Knowledge and Skills to Society and Environment				
Apply knowledge and skills to make connections between science and technology to society and the environment	Makes connections with limited effectiveness	Makes connections with some effectiveness	Makes connections with considerable effectiveness	Makes connections with a high degree of effectiveness
Propose action plans to address problems relating to science and technology, society, and environment	Proposes action plans with limited effectiveness	Proposes action plans with some effectiveness	Proposes action plans with considerable effectiveness	Proposes action plans with a high degree of effectiveness

STUDENT SELF-ASSESSMENT RUBRIC

Name: _____ Date: _____

Put a checkmark in the box that best **describes you.**

Expectations	Need to do Better	Sometimes	Almost Always	Always
I am a good listener.				
I followed the directions.				
I stayed on task and finished on time.				
I remembered safety.				
My writing is neat.				
My pictures are neat and coloured.				
I reported the results of my experiment.				
I discussed the results of my experiment.				
I know what I am good at.				
I know what I need to work on.				

I liked _____

_____

I learned _____

_____

I want to learn more about _____

_____

OTM-2171   ISBN: 978-1-4877-0199-4 © On The Mark Press

INTRODUCTION

The lessons and experiments in this book fall under 5 main topics that relate to the Alberta curriculum for Grade 2 Science – Topic A: Exploring Liquids, Topic B: Buoyancy and Boats, Topic C: Magnetism, Topic D: Hot and Cold Temperatures and Topic E: Small Crawling and Flying Animals. In each lesson you will find teacher notes designed to provide you guidance with the learning intentions, the success criteria, materials needed, a lesson outline, as well as provide some insight on what results to expect when the experiments are conducted. Suggestions for differentiation or accommodation are also included so that all students can be successful in the learning environment.

Throughout the experiments, the scientific method is used. The scientific method is an investigative process which follows five steps to guide students to discover if evidence supports a hypothesis.

1. **Consider a question to investigate.** For each experiment, a question is provided for students to consider. For example, "Does direct sunlight affect water temperature?"

2. **Predict what you think will happen.** A hypothesis is an educated guess about the answer to the question being investigated. For example, "I believe that direct sunlight warms up the temperature of water." A group discussion is ideal at this point.

3. **Create a plan or procedure to investigate the hypothesis.** The plan will include a list of materials and a list of steps to follow. It forms the "experiment."

4. **Record all the observations of the investigation.** Results may be recorded in written, table, or picture form.

5. **Draw a conclusion.** Do the results support the hypothesis? Encourage students to share their conclusions with their classmates, or in a large group discussion format.

ASSESSMENT & EVALUATION:

Students can complete the Student Self-Assessment Rubric in order to determine their own strengths and areas for improvement. Assessment can be determined by observation of student participation in the investigation process. The classroom teacher can refer to the Teacher Assessment Rubric and complete it for each student to determine if the success criteria outlined in the lesson plan has been achieved. Determining an overall level of success for evaluation purposes can be done by viewing each student's rubric to see what level of achievement predominantly appears throughout the rubric.

MEETING YOUR STUDENTS' NEEDS:

Depending on the needs of the students in your class, the teacher may want to scan any Teacher Notes into a digital format. By doing this, no matter the reading abilities of your students, they will be able to access the information of the text. When appropriate, teachers are also encouraged to allow students to collaborate on as many activities possible. This allows all students to be successful without modifying the text significantly.

TOPIC A: EXPLORING LIQUIDS CHARACTERISTICS OF LIQUIDS

LEARNING INTENTIONS:

Students will learn about the properties of liquids.

SUCCESS CRITERIA:

- identify different liquids by their characteristics
- record observations in a chart and use the information to identify and describe liquids
- make conclusions about common properties or characteristics of liquids

MATERIALS NEEDED:

- plastic cups labelled A, B, C, D, E
- a variety of liquids (such as water, milk, vinegar, juice (orange and grape), syrup, ketchup, honey, glue, liquid soap, olive oil, peanut butter and molasses)
- a cookie tray, a plastic tray or bag, newspaper, white paper, paper towel
- a very low ramp to test liquid flow
- a popsicle stick or stir stick or spoon
- a fan, a heater and a hair dryer
- cleaning and safety materials if needed
- copy of *Properties of Liquids* Worksheet for each student or group
- copy of *Describing Liquids* Worksheet for each student or group
- copy of *Comparing Liquids* Worksheet for each student or group
- copy of *Observing Liquids* Worksheet for each student or group
- copy of *When Things Get Wet* Worksheet for each student or group
- pencils, pencil crayons, clipboards for students as needed
- cleaning products as needed

PROCEDURE:

*The following worksheets may be used as one long lesson, as separate shorter lessons or as science lab learning stations.**

1. Discuss with students the meaning of liquids. Lead students in a brainstorming activity or discussion about the different liquids they know. Give students the Properties of Liquids Worksheet to complete. Come back together as a large group to take up answers.

2. Give students the Describing Liquids Worksheet. Tell students you will demonstrate each liquid, but they must make observations. Pour liquids one at a time (olive oil, peanut butter, glue, syrup, honey, grape juice, ketchup, milk, juice, water) one at a time into a bowl (you may want to use a separate or disposable bowl for the glue). Each time, call upon one or two students to touch each liquid and describe how it feels to the rest of the group. Students will compete the worksheet as each liquid is being explored.

3. Ask students what happens when a liquid touches some other object or material. Lead a discussion on how some things will absorb liquids while others will not. Give students the Comparing Liquids Worksheet (front and back) and read through the introduction, materials needed and What To Do sections. As a large group or in small student groups, conduct the experiment. Students will make predictions, record their predictions and observe that will happen when liquid is poured on the different materials. Conduct the experiment by slowly pouring a spot of water and a spot of oil on the paper towel, white paper, newspaper and cookie tray. Students will fill in the lines and the chart on the back of the worksheet.

4. Lead students in a discussion about how we know what liquid is what we think it is. Some liquids, for example, have a smell, a texture or feel, or a certain colour. Explain to students how it could be important to test and observe liquids before us-

OTM-2171   ISBN: 978-1-4877-0199-4 © On The Mark Press

ing them (for example, vinegar and water look the same. How can we tell them apart?) Give students the Observing Liquid Worksheet. Read through each section with the students. Students will work through the steps and observe each liquid in the cups in order to fill in the chart and the blanks on the worksheet.

5. Ask the students if they have ever been outside in the rain. What happens to their clothes? (Or role-play an example, where a shirt absorbs a small amount of rain or a drink). Discuss with the students the ways things can get dry (for example, what does wind do, or heat do, to wet things?). Give students the When Things Get Wet Worksheet and read through the introductory question, the What I Need, the What I Do Section and the questions on the worksheet. Demonstrate the procedure as needed by drying a sample piece of fabric with the fan, the heater and the hair dryer.

DIFFERENTIATION

Slower learners may benefit from working with a peer to identify and compare liquids the worksheets. Other accommodations may include circling identified liquids rather than colouring them or working through the worksheets verbally instead of in written form.

For enrichment, faster learners could look through magazines and cut out pictures of liquids and glue them on paper to make a collage. Students would also practise the spelling of names of liquids and explore common uses for each liquid.

Name: _____

Properties of Liquids

Liquids are materials that change shape. They take the shape of their container. We can pour liquids. Water and juice are some examples of liquids. Colour the liquids below.

OTM-2171   ISBN: 978-1-4877-0199-4 © On The Mark Press

Name: _____

Describing Liquids

There are a lot of words used to describe the properties of liquids. Draw a line from each word to at least one liquid it describes. Each word can describe more than one liquid.

olive oil

sticky

water

thick

peanut butter

grape juice

thin

glue

slick

ketchup

clear

maple syrup

slippery

milk

oozing

honey

orange juice

runny

Comparing Liquids

We know that liquids have different properties. Some liquids pour faster than others, and some are thicker than others. Let's take a closer look at how well liquids can be absorbed by different materials.

You'll need:

- water
- oil
- paper towel
- white paper
- newspaper
- cookie tray

What to do:

1. Make predictions about what will happen when a liquid is poured on a paper towel, white paper, newspaper, and a cookie tray. Record them on Worksheet 6.

2. Slowly pour a small amount of each liquid onto the paper towel, white paper, newspaper, and the cookie tray.

3. What happened to the water? What happened to the oil? Record your observations.

4. Make conclusions about what you have observed.

OTM-2171   ISBN: 978-1-4877-0199-4 © On The Mark Press

Name: _____

Let's Predict

What do you think will happen when you pour a liquid on:

...the paper towel? _____

...the white paper? _____

...the newspaper? _____

...the cookie tray? _____

Let's Observe

	What happened to the water?	What happened to the oil?
paper towel		
white paper		
news- paper		
cookie tray		

Let's Conclude

What is the same for water and oil?

_____

What is different between water and oil?

_____

What was the best material to hold in the liquids?

_____

Observing Liquids Worksheet

What are the common properties of liquids? How can I investigate the properties of liquids?

You'll need:

- plastic cups labelled A, B, C, D, E
- a variety of liquids (such as water, milk, vinegar, juice, syrup, ketchup, mustard, liquid soap, oil or molasses)
- a tray
- a very low ramp to test liquid flow
- a popsicle stick or stir stick or spoon

What to do:

1. Choose one cup of liquid at a time.

2. Using your sense of sight, describe what the liquid looks like. Think of colours and if the liquid is clear or not clear.

3. Using your sense of smell, describe what the liquid smells like. Think of words like good or bad, fruity or flowery and does the liquid have a smell at all?

4. Using your sense of touch, describe what the liquid feels like. Think of words like thick or thin and does it stay on your skin or does it drip clean away?

5. Pour a small amount on the liquid ramp. Does it flow quickly or slowly? Does it flow by spreading out or does it stay together?

OTM-2171   ISBN: 978-1-4877-0199-4 © On The Mark Press

6. Look at the surface of the liquid when it is calm. Think of words to describe it. Would you say it is smooth or rough or wavy?

7. Dip a spoon or stir stick into the liquid and lift it up above the cup. Watch the drops fall. What is the shape of the drops?

5. Write your observations in the chart below and answer the questions on the worksheet.

Sense	Liquid A	Liquid B	Liquid C	Liquid D	Liquid E
Look					
Smell					
Feel					
Flow on the Ramp					
Surface of Liquid					
Shape of Drops					
I think it is…					

What liquid is your favourite? _____

Why was it your favourite? _____

Which liquid do you think is water? _____

When water is calm, what does its surface look like?

_____

Name: _____

When Things Get Wet Worksheet

What are some things I can do when something gets wet? What helps to dry things quickly?

You'll need:

- pieces of fabric, either clothing or small cut samples
- a fan, a heater (or a very sunny day) and a hair dryer
- plastic cups with water
- string for a clothesline or a rack to hang the fabric
- clothes pins or something to hold the fabric in place when hanging

What to do:

1. Take one piece of fabric and dip a corner into the water.

2. Touch the wet fabric to test how it feels when it is wet.

3. Hold the fabric in front of a fan, or hang it up in front of a fan for 1 to 2 minutes.

4. After that time, touch the fabric again. Does it feel very dry, dry, wet or very wet?

5. Do steps 1 and 2 with another piece of fabric.

6. Hold this piece of fabric up to a heater, or hang it up in front of a heater, for 1 to 2 minutes.

7. After that time, touch the fabric again. Does it feel very dry, dry, wet or very wet?

8. Do Steps 1 and 2 again with another piece of fabric.

OTM-2171   ISBN: 978-1-4877-0199-4 © On The Mark Press

9. Hold the fabric in front of a hair dryer, or hang it up in front of a hair dryer for 1 to 2 minutes.

10. After that time, touch the fabric again. Does it feel very dry, dry, wet or very wet?

11. Fill in the chart with your observations and answer the questions below

What I Used to Dry	Before	After
Fan		
Heater		
Hair Dryer		

The fan uses wind. Did the wind from the fan dry the fabric? _____

The heater uses heat. Did the heat from the heater dry the fabric? _____

The hair dryer uses both wind and heat. Did this dry the fabric? _____

What method dried the fabric the best? _____

Bonus: Imagine if you rolled the fabric up and squeezed it. Do you think this would help to dry the fabric even faster? _____

Why? _____

_____

OTM-2171   ISBN: 978-1-4877-0199-4 © On The Mark Press

WHAT IS WATER?

LEARNING INTENTION:

Students will explore water through their senses and learn about the three states of water.

SUCCESS CRITERIA:

- explore water using the five senses
- describe how water looks, smells, tastes, feels, and sounds
- make predictions on how water can change
- record observations in words and drawings
- make conclusions and connections about water and its importance in the environment
- provide evidence of its existence in our environment in its three different states

MATERIALS NEEDED:

- a copy of *Exploring Water* Worksheet 1, 2, 3, and 4 for each student
- a copy of *Three States of Water* Worksheet 5, 6, 7, and 8 for each student
- a beaker of water, 2 small cups, a shallow pan, a magnifying glass (for each pair)
- an ice cube tray, a large baking dish (for each pair of students)
- a kettle of water, a cold plate, an oven mitt
- paper towels, chart paper, markers, pencils, pencil crayons

PROCEDURE:

* **This lesson can be done as one long lesson, or be divided into shorter lessons.**

1. Explain to students that they will do some exploring with water, using their five senses. Give them Worksheets 1, 2, 3, and 4. Read through the question, materials needed, and what to do sections with the students to ensure their understanding. Divide students into pairs and give them the materials to do the exploration. Once they have completed the exploration, they will make some conclusions and connections about water and its importance to us. A follow-up option is to come back as a large group to discuss and record responses on chart paper – Conclusions? Wonders? Importance of Water?

2. Explain to students that they are going to investigate the different states of water. Divide students into groups of two, and give them the materials they need to conduct the *Three States of Water* first experiment. Read through the materials needed and what to do sections on Worksheet 5. Students will make predictions about the outcome of the experiment, make observations through drawings, and make conclusions on Worksheet 6.

3. ***This is a teacher directed experiment.** Read through the question, materials needed, and what to do sections on Worksheet 7 with the students. Students will make a prediction, then the teacher will conduct the experiment. Students will make and record observations and conclusions on Worksheet 8.

DIFFERENTIATION:

Slower learners may benefit by working together in a small group with teacher support to complete written descriptions on Worksheet 2, 3, and 4. An additional accommodation could be to discuss the 'Challenge Question' as a small group, in place of individual written responses.

For enrichment, faster learners could write a poem about water and illustrate it. They could use only one word per sense to describe water. The poem could take a form such as:

Water looks _____.

Water smells _____.

Water tastes _____.

Water feels _____.

Water sounds _____.

OTM-2171   ISBN: 978-1-4877-0199-4 © On The Mark Press

Exploring Water

Question: How would you describe water?

You'll need:

- a small cup
- a shallow pan
- a beaker of water
- a paper towel
- a magnifying glass

What to do:

1. Using the magnifying glass, and your sense of sight, describe how water looks. Record your answer and illustrate what water looks like on Worksheet 2.

2. Use your sense of smell to describe how water smells. Record your answer on Worksheet 2.

3. Pour some water from the beaker into a small cup. Take a drink and use your sense of taste to describe how water tastes. Record your answer on Worksheet 3.

4. Pour some water from the beaker into a shallow pan. Use your sense of touch to describe how water feels. Record your answer on Worksheet 3.

5. Listen carefully as you pour some more water from the beaker into the small cup. Use your sense of hearing to describe how water sounds. Record your answer on Worksheet 3.

7. Make some conclusions about water on Worksheet 4.

8. Answer the challenge question on Worksheet 4.

OTM-2171   ISBN: 978-1-4877-0199-4 © On The Mark Press

Name: _____

Let's Connect

Use your sense of sight to describe how water **looks**.

Water looks _____

_____

_____

_____

An illustration of water:

Use your sense of smell to describe how water **smells**.

Water smells _____

_____

_____

_____

OTM-2171   ISBN: 978-1-4877-0199-4 © On The Mark Press

Name: _____

Use your sense of taste to describe how water **tastes**.

Water tastes _____

_____

_____

_____

Use your sense of touch to describe how water **feels**.

Water feels _____

_____

_____

_____

Use your sense of hearing to describe how water **sounds**.

Water sounds _____

_____

_____

_____

_____

Name: _____

Let's Conclude

What have you discovered about water? What conclusions can you make?

_____

_____

_____

_____

What do you wonder about water?

_____

_____

_____

_____

Let's Connect It!

Challenge Question: Why do you think water is so important to our environment?

_____

_____

_____

_____

_____

OTM-2171   ISBN: 978-1-4877-0199-4 © On The Mark Press

MIXING IT UP!

LEARNING INTENTION:

Students will learn about the interactions of solids and liquids, and how water is interchangeable from a solid, a liquid, and a gas.

SUCCESS CRITERIA:

- identify the different properties of water
- make predictions and investigate how different materials interact
- make observations and record data in a chart
- make conclusions and connections to the environment

MATERIALS NEEDED:

- plastic cups, stir sticks (three per student)
- liquids: water, oil, milk, juice
- solids: salt, sugar, rice, coins, sand
- a copy of *Mixing It Up!* Worksheet 1 and 2 for each student
- ice cube trays (one per two students)
- large baking dishes or pans (one per two students)
- a copy of *Water Detectives* Worksheet 3 and 4 for each student
- a kettle of water, a cold plate, an oven mitt
- a copy of *Making It Rain!* Worksheet 5 and 6 for each student
- a copy of *Provide the Evidence* Worksheet for each student
- pencils

PROCEDURE:

1. Explain to students that they are going to investigate the interactions of solids and liquids. Give students Worksheet 1, and read through the materials needed and what to do sections with them. Give them the materials needed to conduct the experiment. Students will make predictions, investigate how different materials interact, share and compare an observation with a partner, and make conclusions on Worksheet 2.

2. Explain to students that they are going to investigate the different states of water. Divide students into groups of two, and give them the materials they need to conduct the *Water Detectives* experiment. Read through the materials needed and what to do sections on Worksheet 3. Students will make predictions about the outcome of the experiment, make observations through drawings, and make conclusions on Worksheet 4.

3. *This is a teacher directed experiment. Read through the question, materials needed, and what to do sections on Worksheet 5 with the students. Students will make a prediction, then the teacher will conduct the experiment. Students will make and record observations on Worksheet 6.

4. Give students *Provide the Evidence* Worksheet 7. They will use pictures and/or words to show how water exists in our environment in its three different states. Sample responses:

   Solid – ice, snow, hail, sleet, frost
   Liquid – rain, dew
   Gas – fog, water vapor

DIFFERENTIATION:

Slower learners may benefit by working in a small group of three to complete the "Mixing It Up!" experiment. Each group member could be assigned one mixing task then share the results with other members in the group in order to record all results needed to complete the worksheet.

For enrichment, faster learners could write about a solid, a liquid, and a gas, that they would like to know more about. With access to the internet, they could research it.

Name: _____

Mixing It Up!

We mix things together all of the time. What happens when we mix different solids together? What happens when we mix different liquids together? What happens when we mix different solids and liquids together? Let's investigate!

You'll need:

- 3 plastic cups
- 3 stir sticks
- liquids to choose from: water, oil, milk, juice
- solids to choose from: salt, sugar, rice, coins, sand

What to do:

1. Choose two liquids and write down their names in the first column on Worksheet 2.

2. Make a prediction about whether the materials you chose will mix. Record it on your worksheet.

3. Pour the materials into a cup. Stir them with a stick.

4. Did they mix? Record your observations.

5. Repeat Steps 1 to 4 using two solids.

6. Repeat Steps 1 to 4 using a liquid and a solid.

7. Share one of your observations with a friend.

8. Make conclusions about what you observed.

OTM-2171   ISBN: 978-1-4877-0199-4 © On The Mark Press

Name: _____

Let's Observe

Materials Mixed	Predict: Will it mix?	Observe: Did it mix?
Liquid and Liquid		
Solid and Solid		
Liquid and Solid		

Share one of your observations with a partner. Record one of your partner's observations in the chart below.

My partner mixed...	This is what happened...

Let's Conclude

Did any of your materials dissolve? Explain.

_____

_____

Did any of your materials float? Explain.

_____

_____

Did any of your materials sink? Explain.

_____

_____

Name: _____

Water Detectives

Is water always a liquid? No! It can be a liquid, a solid, or a gas. Let's investigate this idea!

You'll need:

- ice cube tray
- water
- large baking dish

What to do:

(Part 1)

1. Pour water into the ice cube tray. Your teacher will put the ice cube tray in the freezer for 24 hours.

2. What do you think will happen? Record your prediction on Worksheet 4.

3. After 24 hours, take the tray out of the freezer. Make a sculpture with your ice cubes in the deep baking dish. Draw a picture of your ice sculpture.

(Part 2)

4. Leave your ice sculpture in the classroom. What do you think will happen to the sculpture? Record your prediction.

5. What happens? Draw a picture.

6. Make conclusions about what you observed.

OTM-2171   ISBN: 978-1-4877-0199-4 © On The Mark Press

Name: _____

Let's Predict

(Part 1)

What will happen when you put the water in the freezer?

_____

_____

(Part 2)

What will happen to your ice sculpture after a few hours?

_____

_____

Let's Observe

Draw a picture of your ice sculpture.

Draw a picture of your ice sculpture after a few hours.

Let's Conclude

What conditions does water need to turn to ice?

_____

What conditions does ice need to turn to water?

_____

Why do you think it rains in the summer and snows in the winter? _____

_____

Name: _____

Making It Rain!

Question: Can water turn into a gas?

You'll need:

- a kettle full of water
- a cold plate
- an oven mitt
- access to an electrical plug

What to do:

1. Make a prediction about the answer to the question.

2. Your teacher will plug in the kettle until the steam begins to rise. Record your observations.

3. Your teacher will hold the plate over the steam. Record your observations.

4. Draw a picture of what you saw happen.

OTM-2171   ISBN: 978-1-4877-0199-4 © On The Mark Press

Name: _____

Let's Predict

Can water turn into gas?

_____

Let's Observe

What happens to the water inside the kettle?

_____

_____

What happens to the steam when it hits the cold plate?

_____

_____

Draw a picture of what you saw happen:

Let's Conclude

When it is hot, water evaporates from lakes. When it rains, water falls back down. Colour the picture of rain.

OTM-2171   ISBN: 978-1-4877-0199-4 © On The Mark Press

Name: _____

Provide the Evidence!

You have learned that water exists in three different states. These states are as a solid, as a liquid, and as a gas.

Using pictures or words, give examples of these three states of water in our environment.

Water as a **solid**:

Water as a **liquid**:

Water as a **gas**:

OTM-2171   ISBN: 978-1-4877-0199-4 © On The Mark Press

THE WATER CYCLE

LEARNING INTENTION:

Students will learn about the water cycle and factors that affect evaporation rate.

SUCCESS CRITERIA:

- identify and describe the water cycle in our natural environment
- make predictions and investigate factors that affect evaporation
- gather data and record observations in a chart
- make conclusions about evaporation rate
- make a connection to the environment and daily living

MATERIALS NEEDED:

- a copy of *The Water Cycle* Worksheet 1 for each student
- a copy of *Evaporation Exposure!* Worksheets 2, 3, 4, and 5, for each student
- 2 jars (one with lid, other without), a ruler, a measuring cup (for each group of students)
- a few jugs of water
- 2 clothes drying racks, 2 small towels, 2 clothes pins, access to water, a warm windy day
- pencils

PROCEDURE:

* **This lesson can be done as one long lesson, or be divided into two shorter lessons.**

1. Give students Worksheet 1. Read through the information about nature's water cycle. Discuss the concepts of 'condensation', 'precipitation', and 'evaporation' with students to ensure their understanding of how each is created and part of the cycle.

2. In small groups or pairs, students will conduct an experiment to investigate evaporation. Give them Worksheets 2 and 3. Read through the question, materials needed, and what to do sections on Worksheet 2. After making a prediction, students will conduct the experiment and record observations on the first day and again seven days later. They will make a conclusion about water evaporation on the seventh day.

3. Students will conduct another experiment to investigate evaporation. Give them Worksheets 4 and 5. Read through the question, materials needed, and what to do sections on Worksheet 4. After making a prediction, students will conduct the experiment and record observations once the wet towels have been put on to the drying racks and again about three hours later. They will make a conclusion about what can affect water evaporation. They will make a connection by explaining the usefulness of this knowledge to their own daily lives.

DIFFERENTIATION:

Slower learners may benefit by having only to observe and record the changes in the height of the water in the jars for experiment #1 (eliminating step 9 in the 'what to do' section). Whiting out the mL sentences in the observation chart, and the challenge question will lend to less confusion on the page for these learners.

For enrichment, faster learners could make their own illustration of the water cycle, labelling it with information they have learned.

Name: _____

The Water Cycle

You have learned that water exists in three different forms. It can be a liquid, a solid, or a gas.

Now let's learn exactly how these forms of water cycle through our environment.

Condensation happens when water in a gas form, meets cooler air. It will form a cloud in the sky where this gas changes into droplets of water.

These droplets of water, called precipitation, fall back down to Earth. **Precipitation** can be in the form of rain, snow, sleet, or hail. It wets the ground, and fills up rivers, lakes, oceans, and streams.

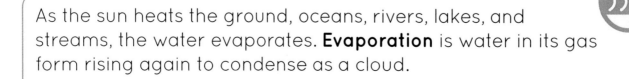

As the sun heats the ground, oceans, rivers, lakes, and streams, the water evaporates. **Evaporation** is water in its gas form rising again to condense as a cloud.

OTM-2171   ISBN: 978-1-4877-0199-4 © On The Mark Press

Name: _____

Evaporation Exposure! #1

Question: Does water need to be exposed to air in order to evaporate?

You'll need:

- 2 jars (one with a lid, the other without)
- a ruler
- a measuring cup
- a jug of water

What to do:

1. Make a prediction about the answer to the question, record it on Worksheet 3.

2. Pour 200mL of water into the measuring cup.

3. Pour the water from the measuring cup into Jar 1.

4. Repeat steps 2 and 3, using Jar 2.

5. Put the ruler into each jar to measure the height of the water in each jar. Record this on Worksheet 3.

6. Put the lid onto Jar 1. Leave Jar 2 exposed to the air.

7. Wait seven days. Repeat step 5.

8. Record your observations about the height of the water in each jar.

9. Pour the water from Jar 1 into the measuring cup, record the amount of water. Repeat this step using Jar 2.

10. Make a conclusion about what you have observed.

OTM-2171   ISBN: 978-1-4877-0199-4 © On The Mark Press

Name: _____

Let's Predict

Does water need to be exposed to air in order to evaporate? Explain your thinking.

_____

_____

_____

Let's Observe

	On the first day...	Seven days later...
Jar 1 (lid)	It had _____ mL of water. It had _____ cm of water.	It had _____ mL of water. It had _____ cm of water.
Jar 2 (no lid)	It had _____ mL of water. It had _____ cm of wate.r	It had _____ mL of wate.r It had _____ cm of water.

Let's Conclude

Was your prediction correct? Explain the results.

_____

_____

_____

Challenge question:

How many millilitres of water evaporated from Jar 2 after seven days? _____

OTM-2171   ISBN: 978-1-4877-0199-4 © On The Mark Press

Evaporation Exposure! #2

Question: How do wind and heat affect evaporation?

You'll need:

- 2 clothes drying racks
- 2 small towels or
- pieces of cloth
- 2 clothes pins
- access to water
- a warm windy day

What to do:

1. Make a prediction about the answer to the question. Record it on Worksheet 5.

2. Soak the two small towels in water, and ring them out.

3. Hang one of the towels on a drying rack inside the classroom (away from open windows and the sun).

4. How does this towel feel? Record your observations on Worksheet 5.

5. Hang the other towel on the other drying rack, secure it with clothes pins. Place this rack outside in the sun, in the warm wind.

6. How does this towel feel? Record your observations on Worksheet 5.

7. Wait about three hours. How do both towels feel now? Record your observations on Worksheet 5.

8. Make a conclusion about what you have observed.

Name: _____

Let's Predict

How do you predict that wind and heat affect evaporation?

_____

_____

Let's Observe

	At first ...	Three hours later...
Towel 1 (inside)	It felt _____ _____ _____	It felt _____ _____ _____
Towel 2 (outside)	It felt _____ _____ _____	It felt _____ _____ _____

Let's Conclude

Was your prediction correct? Explain the results.

_____

_____

Let's Connect It!

How is this information useful to you?

_____

_____

OTM-2171   ISBN: 978-1-4877-0199-4 © On The Mark Press

WATERPROOF OR WATERLOGGED?

LEARNING INTENTION:

Students will learn about materials that absorb water and the materials that repel water.

SUCCESS CRITERIA:

- identify the types of materials that absorb water and types that repel water
- make predictions and investigate which materials are waterproof
- gather data and record observations in a chart
- make conclusions about materials that are waterproof

MATERIALS NEEDED:

- aluminum foil, plastic wrap, newspaper, tissue paper, a cotton cloth, 5 cotton balls, a cup of water
- duct tape
- a copy of *Waterproof Or Waterlogged?* Worksheet 1 and 2 for each student
- a copy of *Keeping Dry* Worksheet 3, for each student
- pencils

PROCEDURE:

1. Discuss with students the meaning of "waterproof" and "waterlogged", to ensure their understanding that waterproof materials repel water and waterlogged materials absorb water. Have students name some objects that they think may be waterproof.

2. Explain to students the purpose of the experiment by reading through the question, materials needed, and what to do sections on Worksheet 1.

3. Instruct students to make predictions for each material listed in the chart, as to whether they are waterproof or not. Conduct the experiment by immersing a cotton ball wrapped in each material into the cup of water. Students will record the results in the chart, and make conclusions at the end of the experiment.

4. Give each student a copy of the *Keeping Dry* crossword puzzle to complete

DIFFERENTIATION:

Slower learners may benefit by pairing up with a peer in order to review and discuss the vocabulary in the Word Box, which would assist them when completing the crossword puzzle. Completing the crossword with a peer could be a further accommodation.

For enrichment, faster learners can draw a design of a piece of waterproof clothing that they could use to stay dry while out in a wet environment. This could be done on the reverse side of the crossword worksheet.

Waterproof or Waterlogged?

Raincoats keep us dry on rainy days. They are made of a special waterproof material. This means the material will repel or push off the water so the water cannot soak through.

Umbrellas, boots, gloves, and snow pants are some other things made of waterproof materials. Let's conduct an experiment to test what other materials are waterproof.

Question: Which materials are waterproof?

You'll need:

- aluminum foil
- plastic wrap
- newspaper
- cotton cloth
- tissue paper
- 5 cotton balls
- a cup of water
- duct tape

What to do:

1. Make predictions about the answer to the question. Record them on Worksheet 2.

2. Place a cotton ball in each material in the list above. Be sure to wrap it up well. Seal each with a piece of duct tape.

3. Put the material into the cup of water.

4. Pull it out of the water, unwrap the material to see if the cotton ball is wet or dry.

5. Record the results for each material in the chart.

OTM-2171   ISBN: 978-1-4877-0199-4 © On The Mark Press

Name: _____

Let's Predict

What will be the answer to your question? Record your predictions in the chart below. Then conduct the experiment.

Let's Observe

What happened during the experiment?

Material	Prediction	Result
newspaper		
aluminum foil		
tissue paper		
plastic wrap		
cotton cloth		

Let's Conclude

The materials that absorbed water were

_____

_____

The materials that repelled water were

_____

_____

What material would you choose to make a raincoat out of? _____

Name: _____

Keeping Dry!

Read the clues and use words from the Word Box to help you fill in the answers to this crossword puzzle.

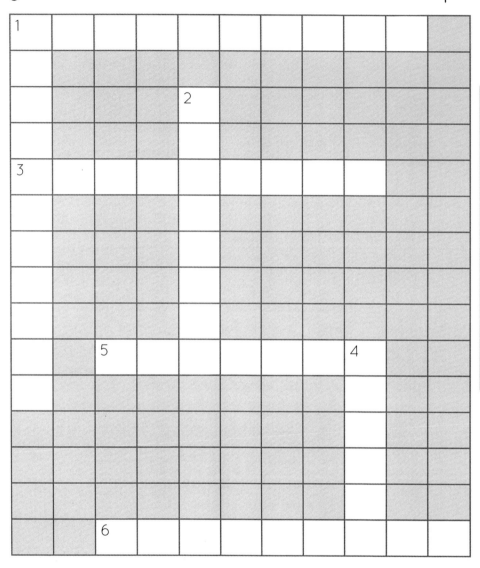

Word Box

plastic

umbrella

rain boots

waterlogged

waterproof

cotton

snow pants

ACROSS

1. This means to repel water.

3. We wear these on our feet to stay dry.

5. This material is waterproof.

6. We wear these in the snow to stay dry.

DOWN

1. This means to absorb water.

2. We carry this to stay dry in the rain.

4. This material absorbs water.

OTM-2171 ISBN: 978-1-4877-0199-4 © On The Mark Press

NEEDING AND USING WATER

LEARNING INTENTION:

Students will learn about where water comes from, the water treatment process, their families' usage of water, and ways to reduce water consumption.

SUCCESS CRITERIA:

- determine where water is found, explain why it is needed, and how it is used
- describe the water treatment process
- recognize the amount of water required to do some daily living activities
- make observations and record results about your family's use of water
- make conclusions about water usage
- make connections to the environment as you look at ways to reduce water usage

MATERIALS NEEDED:

- a copy of *Thinking About Water* Worksheet 1 for each student
- a copy of *Illustrating Your Thinking* Worksheet 2 for each student
- a copy of *How Do We Get It?* Worksheet 3 for each student
- a copy of *How Much Water Do You Use?* Worksheet 4, 5, 6, and 7 for each student
- a copy of *Reduce Your Use!* Worksheet 8 for each student
- calculators
- chart paper, markers, pencils, pencil crayons

PROCEDURE:

* **This lesson can be done as one long lesson, or be divided into four shorter lessons.**

Items 1, 2, 3, 5, 6, 7, and 8 can be done at school. Item 4 contains a homework component.

1. Divide students into pairs. Give them Worksheet 1 to complete. An option upon completion is to come back together as a large group to discuss students' ideas to the questions, recording responses on chart paper. Give students Worksheet 2 to complete.

2. Give students Worksheet 3. Read and discuss as a large group to ensure students' understanding of the information about the water treatment process. An option is to put the Worksheet on an overhead projector or use a document camera to display the information as it is discussed as a group.

3. Give students Worksheet 4. Read and discuss the information as a large group to ensure students' understanding of the information about how we use water in our daily lives, and how much of it we use. Students will complete the question at the bottom of the Worksheet, individually, or as a 'turn and talk' exercise with a peer.

4. Explain to students that they will conduct an investigation about their family's water usage around the house. Give them Worksheet 5. Explain that they will use this chart at home to record how many times water is used by family members in their house. A class discussion about how else water is used in the home may be of benefit, as this may provide guidance to students' choices of 'other' in the chart.

5. After students have had 2 days to complete Worksheet 5, ask them to return it to school so that they can assess the data that they collected. Give students Worksheet 6. Students will count up the number of squares they coloured in for each activity on Worksheet 5 and record this on Worksheet 6. Referring back to Worksheet 4, students will record the litres of water required to do each of the activities listed on Worksheet 6. Using a calculator, they will multiply the number

of times the activity was done by the number of litres of water it takes to do it each time. Then they will record this information in the chart on Worksheet 6. (It may be beneficial to do this activity in small groups, one group at a time, so that teacher direction and support can be given to each student as they calculate their data.)

6. Divide students into pairs and give them Worksheet 7. Students will engage in a 'Share and Compare' exercise as they analyze, assess, and compare the data that they collected about their families' water usage. An option is to come back together as a large group to discuss their findings and ideas.

7. Have a large group discussion about the following ideas:

   • What does it mean to use water excessively?

   • What does it mean to use water efficiently?

   • How might our water usage change if we had to carry our water into our homes instead of just turning on the tap?

8. Divide students into pairs and give them Worksheet 8. Students will engage in a 'Think-Pair-Share' activity in which, they will discuss and respond about responsible water use. An option is to come back together as a large group to discuss and record their ideas on chart paper.

DIFFERENTIATION:

Slower learners may benefit by only recording their own thinking on Worksheets 1, 7, and 8; and only listening to their partner's responses that are shared.

For enrichment, faster learners could create a bar graph that depicts their data collected for their family's water usage. An alternative or additional enrichment activity for these learners would be to have them create a poster to illustrate ways we could reduce our water usage.

OTM-2171   ISBN: 978-1-4877-0199-4 © On The Mark Press

Name: _____

Thinking About Water

Think, Pair, Share

With a partner, do some thinking and sharing about the questions below. Record your ideas in the chart.

"Where is water found in the environment?"

My Thinking	My Partner's Thinking
_____	_____
_____	_____
_____	_____

"Why do living things need water?"

My Thinking	My Partner's Thinking
_____	_____
_____	_____
_____	_____

"How is water used?"

My Thinking	My Partner's Thinking
_____	_____
_____	_____
_____	_____

OTM-2171   ISBN: 978-1-4877-0199-4 © On The Mark Press

Name: _____

Illustrating Your Thinking

Use illustrations to show where water can be found in our environment.

Illustrate some of the living things that depend on water.

This is how I use water in my daily life.

OTM-2171   ISBN: 978-1-4877-0199-4 © On The Mark Press

Name: _____

How Do We Get It?

You know that you need water. You know that you use it every day. Have you ever wondered how you get it to your tap? Let's take a look at the water treatment process!

Water is brought into a water treatment plant from a natural source like a lake. Once the water is in the plant, it goes through many steps to be cleaned and ready to come out of your tap.

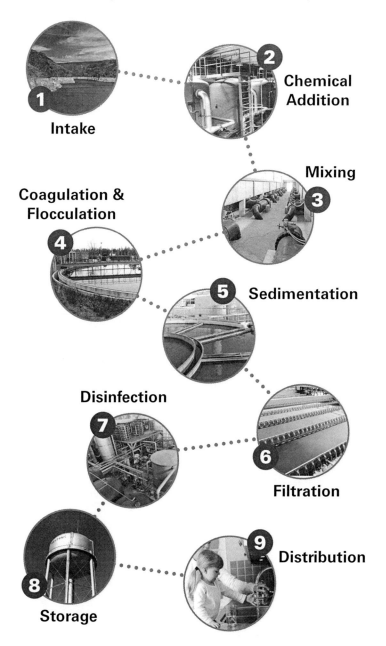

1 Intake

2 Chemical Addition

3 Mixing

4 Coagulation & Flocculation

5 Sedimentation

6 Filtration

7 Disinfection

8 Storage

9 Distribution

Steps:

1. Water is taken from a natural source, like a lake.

2. Chemicals are added to the water to kill germs.

3. The chemicals and the water get mixed together.

4. Particles stick together to form flocculants or "floc."

5. The water and floc go into a basin. The floc sinks to the bottom and is taken out of the water.

6. Water flows through filters.

7. More chemicals are added to kill any other germs.

8. Water is stored in a tank.

9. Water is taken to houses and businesses.

OTM-2171   ISBN: 978-1-4877-0199-4 © On The Mark Press

Name: _____

How Much Water Do You Use?

In the chart below are some examples of how we use water in our homes and how much of it we use.

Daily activities that use water	Average amount of litres it takes
Brushing your teeth	1 litre
Flushing the toilet	6 litres
Washing your hands and face	7 litres
Taking a bath	80 litres
Taking a 5 minute shower	35 litres
Running the dishwasher	20 litres
Washing a sink of dishes by hand	30 litres
Using the clothes washing machine	70 litres
Drinking water per day	1 litre

What are some other ways that we use water in and around our homes?

_____

_____

_____

OTM-2171   ISBN: 978-1-4877-0199-4 © On The Mark Press

Name: _____

Let's Collect

Record how often water is used in your home by your family for 2 days. Colour in one square in the correct section each time it is used.

Brushing teeth	
Flushing toilet	
Washing hands or face	
Taking a bath	
Taking a 5 minute shower	
Running the dishwasher	
Hand washing a sink of dishes	
Using the washing machine	
Drinking water per person	
Other? _____	
Other? _____	

Let's Calculate

Calculate how much water, was used by your family for 2 days. You will need to look at the information on Worksheet 4 and the data that you collected on Worksheet 5.

Activities using water	Number of squares coloured in the chart	Number of litres of water it takes to do the activity *one time*	Amount of litres of water used to do the activity over *2 days*
Brushing teeth			
Flushing toilet			
Washing hands or face			
Taking a bath			
Taking a 5 minute shower			
Running the dishwasher			
Hand washing a sink of dishes			
Using the washing machine (clothes)			
Drinking water per person			
Other? _____			
Other? _____			

OTM-2171   ISBN: 978-1-4877-0199-4 © On The Mark Press

Share and Compare

Working with a partner, compare your final numbers in the chart on Worksheet 6.

Talk about how many litres of water each of your families used over 2 days. Use the questions in the chart to help you compare.

Questions	My Family	My Partner's Family
In which activity did your family use the most water?		
In which activity did your family use the least water?		
What surprised you about your family's use of water?		
What could your family learn from the data that you collected?		

Name: _____

Reduce Your Use!

Think, Pair, Share

With a partner, do some thinking and sharing about the questions below. Record your ideas in the chart.

"Why shouldn't we waste water?"

My Thinking	My Partner's Thinking
_____	_____
_____	_____
_____	_____

"How is your family already showing smart use of water?"

My Thinking	My Partner's Thinking
_____	_____
_____	_____
_____	_____

"What are some other ways your family could save water?"

My Thinking	My Partner's Thinking
_____	_____
_____	_____
_____	_____

OTM-2171   ISBN: 978-1-4877-0199-4 © On The Mark Press

TOPIC B: BUOYANCY AND BOATS: SINK OR FLOAT?

LEARNING INTENTION:

Students will learn about the buoyancy of different materials, and use this knowledge to design and build a floater.

SUCCESS CRITERIA:

- identify the types of materials that float in water and those that sink in water
- make predictions, investigate buoyancy of materials, and record observations in a chart
- make a plan to construct a floater, make and record observations of the final product by comparing and sharing similarities, benefits of the design, and testing for buoyancy
- make conclusions about materials that are buoyant

MATERIALS NEEDED:

- a clear plastic tub of water (one per group of students)
- a coin, a straw, a tennis ball, a feather, a button, a jar lid, unpopped corn kernels, popped corn
- kernels, a deflated balloon, and inflated balloon (per group of students)
- a copy of *Sink or Float?* Worksheet 1 and 2 for each student
- egg cartons, Styrofoam trays or plates, plastic bottles and milk jugs, cardboard tubes, tart tins, sponges, margarine tubs or other plastic containers, glue, string, tape, elastics, butterfly fasteners, paper clips, aluminum foil, toothpicks, popsicle sticks, scissors, hole puncher, coins, marbles, washers, ping pong balls
- a copy of *Building a Floater* Worksheet 3 and 4 for each student
- pencils, markers

PROCEDURE:

1. As a large group, or in small groups, conduct an experiment to explore which materials float. Give students Worksheets 1 and 2, and explain the purpose of the experiment by reading through the question, materials needed, and what to do sections on Worksheet 1. Instruct students to make predictions for each material listed in the chart, as to whether they will float in the water or not. Students will conduct the experiment by putting each material into the tub of water. They will record the results in the chart, and make conclusions at the end of the experiment.

(Students should conclude that the straw, tennis ball, feather, jar lid, popped corn kernels, and inflated balloon float because they have space to hold air, which displaces the water they are floating in)

2. Explain to students that they are going to plan, design, and construct a floater made out of the recyclable materials. They will compare their final product with a partner and make some observations. Give each student a copy of the *Building a Floater* Worksheets 3 and 4 to complete, as they begin to construct their floater. Once floaters are complete, students will test the buoyancy of their designs by putting them in the plastic tubs of water.

DIFFERENTIATION:

Slower learners may benefit by working with a peer in order to choose materials, and make a plan and design for a floater. Assembly and testing of the floater could then be completed as team or independently.

For enrichment, faster learners could experiment further with the materials in experiment 1 by altering a floating object so that it will sink and alter a non-floating object so that it will float.

Name: _____

Sink or Float?

Some solids are heavy and some are light in weight. Do heavy objects always sink in water? Do light objects always float? Let's explore these questions by experimenting with some materials.

Question: Which materials float?

You'll need:

- a clear plastic tub full of water
- a coin
- a tennis ball
- a feather
- a button
- a jar lid
- unpopped corn kernels
- popped corn kernels
- deflated balloon
- inflated balloon

What to do:

1. Make predictions about the answer to the question. Record them on Worksheet 1.

2. Place each material in the list above into the tub of water.

3. Record the results for each material in the chart.

4. Make conclusions about what you have observed.

OTM-2171   ISBN: 978-1-4877-0199-4 © On The Mark Press

Name: _____

Let's Predict

What will be the answer to your question? Record your predictions in the chart below, then conduct the experiment.

Let's Observe

Materials	Predictions	Results
coin		
straw		
tennis ball		
feather		
jar lid		
unpopped corn kernel		
popped corn kernel		
deflated balloon		
inflated balloon		

Let's Conclude

Which materials floated in the water?

_____

Why do you think they were able to float?

_____

_____

Name: _____

Building a Floater

Use what you know to design and construct a floater.

The materials I will use are:

_____   _____

_____   _____

_____   _____

My plan for making the floater is:

1._____

2. _____

3. _____

4. _____

5. _____

Draw a design of what your floater will look like:

Now carry out your plan by gathering the materials you need. Begin your construction!

OTM-2171   ISBN: 978-1-4877-0199-4 © On The Mark Press

Name: _____

Working with a partner, compare your final product. List two things that are the same and two things that are different about your floaters.

Things That Are The Same	Things That Are Different
1. _____ _____ _____ 2. _____ _____ _____	1. _____ _____ _____ 2. _____ _____ _____

Let's Observe

Place your floater into the plastic tub of water to test if it will sink or float. Draw a picture of what happened.

Sink or Float?

My floater _____ .

What might you change about your floater?

_____

_____

_____

SHIPS AHOY!

LEARNING INTENTION:

Students will learn about stability, load carrying, and how watercraft can be propelled.

SUCCESS CRITERIA:

- identify the characteristics of a stable and balanced watercraft
- investigate stability by constructing a watercraft that can carry a load
- investigate propulsion by constructing a watercraft that can move through water
- make and record observations of the final products through drawings of the outcome

MATERIALS NEEDED:

- a clear plastic tub of water (one per group of students)
- Plasticine (a couple of handfuls per student)
- pennies (10 per student)
- a 10 oz. plastic soda bottle, baking soda, toilet paper, vinegar, 3 marbles (per group of students)
- a copy of *Ships Ahoy!* Worksheet 1 and 2 for each student
- pencils, pencil crayons, markers

PROCEDURE:

1. Have a discussion with the students to teach them the concept that a boat floats when it displaces the volume of water that is equal to its mass. For example, a sailboat has a low mass with a large underwater volume. Its hull is wide bodied and deep to displace a large volume of water. A sailboat has a keel which gives it a center of gravity so that it won't turn over and sink. A keel is a beam that runs from the bow to the stern inside the hull. It is like a spine that gives the hull structural strength. Also, making the bottom of the boat thicker than the sides, helps to keep the center of gravity low down in the water. Explain to students that they are going to construct a boat out of Plasticine. Their goal is to make sure the boat floats and is able to carry a load. Give students Worksheet 1. Read through the materials needed and what to do sections with them to ensure their understanding. Give students the materials they need to build their boats and test for load carrying ability. They will record the results of their creation by drawing what they have observed.

2. Explain to students that they are going to work with a partner to construct a boat that can propel through water. Give students Worksheet 2. Read through the materials needed and what to do sections with them to ensure their understanding. Give students the materials they need to construct their boats and test for propulsion. They will record the results of their creation by drawing what they have observed.

DIFFERENTIATION:

Slower learners may benefit by having only to construct a boat out of Plasticine that floats, but eliminating the expectation of ensuring that it can carry the load of pennies.

For enrichment, faster learners could experiment further with the materials in experiment 2 by altering the amount of vinegar and baking soda to see if it has an affect on the speed of their boat.

OTM-2171   ISBN: 978-1-4877-0199-4 © On The Mark Press

Name: _____

Ships Ahoy!

We've learned a lot about buoyancy. Now let's design a boat that can carry a load!

You'll need:

- a couple of handfuls of Plasticine
- 10 pennies
- a clear plastic tub

What to do:

1. Shape the Plasticine to make a boat. Here are a few tips:
    - your boat should have a wide body
    - make the bottom thicker than the sides
    - create a center of gravity by giving your boat a small keel

2. Put your boat into the tub of water to check for buoyancy.

3. Carefully place the pennies, one at a time, onto your boat. Be sure to space them out evenly on your boat.

4. Record your observations by drawing your watercraft.

Name: _____

Ships Ahoy!

We've learned a lot about buoyancy. Now let's design a boat that can move!

You'll need:

- a 10 oz. plastic soda bottle
- a few tablespoons of baking soda
- a clear plastic tub of water
- 4 squares of toilet paper (connected)

- 3 marbles
- 1/2 cup of vinegar

What to do:

1. Put 3 marbles into the soda bottle.

2. Fold the toilet paper squares in half so that it is 2 squares in length. Put a few tablespoons of baking soda on the toilet paper and roll it up. Put it in the soda bottle.

3. Pour 1/2 cup of vinegar into the soda bottle. Place the cap on, twist only once.

4. Place the soda bottle in the tub of water and record your observations by drawing what you have created.

This is a picture of my watercraft propelling through water!

OTM-2171   ISBN: 978-1-4877-0199-4 © On The Mark Press

DESIGNING AND BUILDING A BOAT

LEARNING INTENTION

Students will learn how to apply knowledge from investigations and choose appropriate materials and designs in order to build a model boat.

SUCCESS CRITERIA

- design and develop plans, considering properties of materials in decisions and plans
- make and record observations of the final product by comparing and sharing similarities, differences and benefits of designs.
- describe the purpose behind each part of the objects built.

MATERIALS NEEDED

- a copy of *Building a Boat* Worksheet 1 and 2 for each student
- water
- water trench or bucket, sink and tap, or a clear aquarium of water
- popsicle sticks
- aluminum foil
- plastic or styrofoam dinner plates or trays, small plastic cups, 1-2 ounces
- plastic straws, cut in half for sizing
- optional: foam pool noodle toy (cut into 3-cm thick cylinders), or corks, a 500ml plastic water bottle cut lengthwise in half (like a boat hull)

PROCEDURE

1. Ask students if they have ever been on a boat. Did the boat float on top of the water? What was the boat made of?

2. Demonstrate how certain things float in water and certain things sink or soak up water. For example, take a popsicle stick and ask the class if it will float? Stress vocabulary for the students. What is meant by the word "waterproof"? What is an example of some material that is waterproof? Is this material stiff and rigid or flexible? Can you bend it and change its shape? Does this help it stay above the surface of the water? What would be better for a boat? Then test it by placing it

in a bucket or aquarium of water. Were their predictions right? Repeat this step with a piece of aluminum foil (flat or shaped like a bowl), a crumpled up piece of aluminum foil (to show shape can make things float or not float), a piece of paper (soaking up water, falling apart) a plastic cup standing up or on its side (to show position and balance can make things float or not float) and with foam and plastic items (to show material can make things float or not float).

3. Tell students they will design and build a boat. But the boat must float. They must use materials and shapes that will float. They will have to find ways to keep their boats together. Discuss with the class ways of binding things together (rubber bands, glue, masking tape, other) that may or may not be waterproof. Demonstrate examples if needed, such as tying together many popsicles with a rubber band. Test to see that things float when tied together.

4. With the class, read through the Introductory Question, Materials, What to Do and My Design sections on the Building a Boat Worksheet. Read the What I Observe section of the worksheet so that students know what to look for. When students have completed their design they can build their boat. (If students used certain kinds of glue, you may want to do Step 8 the next day.)

5. Set up a station with a sink and tap or with a bucket and water so that students can test their boats. Allow students to change their design and fix their boats as needed. Give students time to record the observations on the worksheet or communicate their findings in a class discussion.

DIFFERENTIATION

Slower learners may benefit from using a pre-designed or partially designed boat. Their task is then to complete the design and build the boat that floats. Another accommodation may involve boats made of fewer parts.

An extension of this activity could involve students finding ways to propel their boats. How would they modify their design in order to attach a sail, a paddling system, or something else to move the boat.

Building a Boat

If you make a boat, will it float? What will you use to make your boat? Will you use plastic, paper, wood, foam, aluminum foil or something else?

You'll need:

- plastic or styrofoam trays
- small sheets of paper
- popsicle sticks
- aluminum foil
- plastic straws

- rubber bands
- glue, masking tape or stapler
- foam or styrofoam pieces
- a bucket of water or a sink and tap

What to do:

1. Think about the different things that float. Think about what you would like your boat to look like. Will it have a sail? Will it have a place to put things? Will it look like a circle, a rectangle, a triangle or some other shape? Draw a picture of your boat in the picture box on this worksheet.

2. Label each part of your boat—what material will you use? How will it stay together?

3. Build your boat.

4. Test it to see if it will float. Make changes to your boat to make it better.

5. Complete the **What I Observe** part of the worksheet.

OTM-2171 ISBN: 978-1-4877-0199-4 © On The Mark Press

Name: _____

My Design

What is the name that I gave my boat? _____

What did I use to make my boat? _____

_____

_____

Why did I use these materials? _____

_____

_____

What did I use to keep the parts of my boat together?

_____

What I Observe

Did my boat float?   **Yes**   **No**

Did my boat stay together?   **Yes**   **No**

What could I do to make my boat even better?

_____

_____

TOPIC C: MAGNETISM: MAGNETISM

Learning Intention:

Students will learn that magnets attract certain materials, and that same type poles repel while opposite poles attract.

Success Criteria:

- identify the materials that magnets attract
- identify where magnets are used in the environment and in our everyday lives
- describe how magnets attract or repel
- make predictions, gather findings about the magnetism of objects and the polarity of magnets, using charts, drawings, and written descriptions
- make conclusions about magnetism and make connections to the environment

Materials Needed:

- bar magnets with poles identified (one for each student)
- a variety of objects such as paper clips, aluminum foil, safety pins, corks, erasers, spoons, Canadian coins, American coins
- a copy of *Magnetism* Worksheet 1 and 2 for each student
- compasses (one for each student)
- cardboard, blunt needles (wool needles), bar magnets, bowls of water (one for each group)
- a copy of *How Does a Compass Work?* and *Making a Compass* Worksheets 3 and 4 for each student
- a copy of *Attract or Repel?* Worksheet 5 and 6 for each student
- pencils

Procedure:

*This lesson can be done as one long lesson, or can be divided into three shorter lessons.

1. Students will investigate which materials magnets attract. Read through the question, materials needed, and what to do sections on Worksheet 1 with the students. They will make their predictions. Conduct the experiment as a large group or in small groups. Using Worksheet 2, students will record results and make conclusions.

2. Lead students into a discussion/brainstorming activity about where magnets are found and how they are used in our lives. Introduce a compass as an object that helps us find direction. The needle is a magnet that always points to the North Pole. Give students "How Does a Compass Work?" and "Making a Compass" Worksheets 3 and 4, and the materials to do the experiments. Read through the question, materials needed, and what to do sections with the students to ensure understanding. They can work with a partner or independently to conduct the experiments, then make observations and conclusions.

3. Remind students that a bar magnet has a north and a south pole. They will investigate the magnetic force of the poles. Give students *Attract or Repel?* Worksheet 5 and 6, and the materials to do the experiments. They can work with a partner or independently to conduct the experiments, then make observations and conclusions.

Differentiation:

Slower learners may benefit by working together in a small group with teacher direction to conduct the *How Does a Compass Work?* and *Make a Compass* experiments.

For enrichment, faster learners could make up a story with a hero who uses a compass. This could be written and illustrated in a graphic text form.

 OTM-2171   ISBN: 978-1-4877-0199-4 © On The Mark Press

Magnetism

Magnets pick up some objects and not others.

When a magnet picks up an object, the object is called magnetic.

An object that is not picked up by a magnet is not magnetic.

Question: What materials will a magnet pick up?

You'll need:

- small magnets
- a variety of objects such as nails, paper clips, aluminum foil, an eraser, a cork, a spoon, safety pins, Canadian coins, American coins

What to do:

1. Make predictions about the answer to the question. Record them on Worksheet 2.

2. Place all of the objects on a table.

3. Observe and record on your worksheet the objects a magnet can pick up and the objects that it can't pick up.

4. Make conclusions about what you have observed.

Name: _____

Let's Predict

What will be the answer to your question? Record your predictions in the chart below.

Let's Observe

What happened when you tried to pick up the objects with the magnet?

Material	Predictions (Will it pick up?)	Pick Up? (Yes or No)
paper clips		
eraser		
aluminum foil		
cork		
safety pins		
spoon		
Canadian coins		
American coins		

Let's Conclude

Which objects did the magnet attract?

_____

What materials are those objects made of ?

_____

Why is it useful to know which materials magnets attract?

_____

_____

OTM-2171   ISBN: 978-1-4877-0199-4 © On The Mark Press

Name: _____

How Does A Compass Work?

A compass is a tool that tells us the direction of the North Pole. The needle in a compass is a magnet. It always points towards the North Pole.

You'll need:

- a compass
- a piece of paper labelled **north**
- a piece of paper labelled **south**

What to do:

1. Place your compass on your desk. Look at the arrow on your compass.

2. Which wall does it point to? Place the "north" label on this wall. Place the "south" label on the opposite wall.

3. Sit facing the back of the classroom, and try again.

4. Record your observations below.

Let's Observe

What do you notice about the direction the compass points to? _____

Let's Conclude

If you wanted to walk to the North Pole, which direction would you go? _____

If you wanted to walk to the South Pole, which direction would you go? _____

Name: _____

Making a Compass

You'll need:

- a compass
- a blunt needle
- a bar magnet
- cardboard
- a small bowl
- half filled with water

What to do:

1. Cut out a disk from the cardboard.

2. Rub the needle with the magnet 30 times. Rub it in the same direction.

3. Place the cardboard disk in the water. Place the needle on the disk. Where does the needle point?

4. Record your observations by drawing what you've made.

Let's Observe

Draw your needle:

Draw your compass:

Does the needle point in the same direction as the compass?

_____

OTM-2171   ISBN: 978-1-4877-0199-4 © On The Mark Press

Attract Or Repel?

We know that the needle in a compass is like a magnet. We also know that a bar magnet has a north and a south pole.

Question: What will happen when a compass is near a magnet?

You'll need:

- a compass
- a bar magnet

What to do:

1. Place the bar magnet on your desk.

2. Move the compass near the **north** pole of the magnet. Draw what happens.

3. Move the compass near the **south** pole of the magnet. Draw what happens.

4. Make a conclusion about what you have observed.

Name: _____

Let's Observe

When I put the compass near the **north** pole of the magnet, this is what happened.

When I put the compass near the **south** pole of the magnet, this is what happened.

Let's Conclude

Circle your answer.

Opposite type poles: **repel**   **attract**

Same type poles: **repel**   **attract**

OTM-2171   ISBN: 978-1-4877-0199-4 © On The Mark Press

MAGNETIC STRENGTH

LEARNING INTENTION:

Students will learn about the effects of magnetic strength and how magnets move and hold things together.

SUCCESS CRITERIA:

- identify and compare the strength of magnets and the materials which block magnets
- make predictions, gather findings about the magnetic strength of magnets, using charts, drawings, and written descriptions
- make a plan to build a magnetic invention, make and record observations of the final product by comparing and sharing similarities, differences, and benefits of the design
- make conclusions about magnetic strength and make connections to the environment

MATERIALS NEEDED:

- bar magnets, horseshoe magnets, round magnets
- paper clips (a box for each group)
- a copy of *Magnetic Strength* Worksheet 1 and 2 for each student
- a piece of plastic, a thin piece of wood, a piece of metal, a piece of paper or cardboard, a piece of lead wrapped in a plastic sandwich bag (a set for each group)
- a copy of *Can Magnets Be Blocked?* Worksheet 3 and 4 for each student
- egg cartons, Styrofoam trays or plates, cardboard tubes, margarine tubs or other plastic containers, newspaper, cardboard, glue, string, tape, elastics, butterfly fasteners, paper clips, aluminum foil, popsicle sticks, pipe cleaners, scissors, hole puncher
- a copy of *Build a Magnet Invention* Worksheet 5 and 6 for each student
- pencils

PROCEDURE:

*This lesson can be done as one long lesson, or be divided into three shorter lessons.

1. Students will investigate which type of magnet is the strongest. Give students *Magnetic Strength* Worksheet 1 and 2. Read through the question, materials needed, and what to do sections with them. Students will make predictions, conduct the experiment in small groups, record results, and make conclusions based on their observations.

2. Students will investigate which materials block magnets. Give students *Can Magnets Be Blocked?* Worksheet 3 and 4 and the materials to do the experiment. Read through the question, materials needed, and what to do sections with the students. They can work in small groups to conduct the experiment, then make observations and conclusions.

3. Explain to students that they are going to plan, design, and build a magnet invention made out of some recyclable materials. Then, they will compare their final product with a partner and make some observations. Give each student a copy of *Build a Magnet Invention* Worksheet 5 and 6 to complete, as they begin to build their invention. Once all magnet inventions are completed, invite students to present their inventions to the class and share three things about it.

DIFFERENTIATION:

Slower learners may benefit by listing only one thing that is the same and one thing that is different from their partner's invention, and provide only one thing to share with the class about their own invention.

For enrichment, faster learners could build an additional object that uses a magnet that would be useful at home.

Name: _____

Magnetic Strength

Some magnets are stronger than others. We can measure the strength of a magnet by how many paperclips it can pick up.

Question: What type of magnet is the strongest?

You'll need:

- a bar magnet, a horseshoe magnet, a round magnet
- a box of paperclips

What to do:

1. Make predictions about the answer to the question. Record them in the chart on Worksheet 2.

2. Pour the paperclips onto a table. Try to stick as many paper clips as you can on the bar magnet.

3. Count how many paperclips stick to the magnet, record the result in the chart below.

4. Repeat steps 2 and 3 using the horseshoe magnet.

5. Repeat steps 2 and 3 using the round magnet.

6. Make a conclusion about what you have observed.

OTM-2171   ISBN: 978-1-4877-0199-4 © On The Mark Press

Worksheet 2

Name: ______________________________

Let's Predict

What type of magnet is the strongest? Circle your prediction.

bar magnet **horseshoe magnet** **round magnet**

Let's Observe

Record your results in the chart below.

Type of Magnet	Number of Paperclips that Stuck
bar magnet	
horseshow magnet	
round magnet	

Let's Conclude

Which magnet is the strongest?

____

Why do you think it proved to be the strongest?

____

____

____

____

____

Name: _____

Can Magnets Be Blocked?

Question: What materials can block a magnet?

You'll need:

- a horseshoe magnet
- a paperclip
- a piece of plastic
- a piece of wood
- a piece of metal
- a piece of paper or cardboard
- a piece of lead in a plastic sandwich bag

What to do:

1. Make predictions about the answer to the question. Record them on Worksheet 4.

2. Place the magnet on a table.

3. Cover the magnet with a piece of plastic. Place a paperclip on the plastic. Is it attracted to the magnet? Record your results.

4. Repeat step 3 using a piece of wood.

5. Repeat step 3 using a piece of metal.

6. Repeat step 3 using a piece of paper or cardboard.

7. Repeat step 3 using the lead wrapped in a sandwich bag.

8. Make a conclusion about what you have observed.

OTM-2171   ISBN: 978-1-4877-0199-4 © On The Mark Press

Name: _____

Let's Predict

Will the magnet still attract the paperclip? Write "**yes**" or "**no**" in the first column.

Let's Observe

Does the magnet still attract the paperclip? Write "**yes**" or "**no**" in the second column.

Materials Used	Predictions	Results
plastic		
wood		
metal		
paper		
lead		

Let's Conclude

Which material blocks the magnet?

_____

_____

_____

Why is it useful to know which material blocks magnets?

_____

_____

_____

OTM-2171   ISBN: 978-1-4877-0199-4 © On The Mark Press

Name: _____

Build A Magnet Invention

Invent an object that uses magnets that would be useful for a student.

What is your idea?

_____

_____

The materials I will use are:

_____   _____

_____   _____

_____   _____

My plan for building my magnet invention is:

1. _____

2. _____

3. _____

4. _____

5. _____

Draw what your magnet invention will look like:

Now gather your materials and begin construction!

OTM-2171   ISBN: 978-1-4877-0199-4 © On The Mark Press

Name: _____

Share and Compare

Working with a partner, compare your final product.

List two things that are the same and two things that are different about your inventions.

Things That Are The Same	Things That Are Different
1. _____ _____ _____ 2. _____ _____ _____	1. _____ _____ _____ 2. _____ _____ _____

Three things I would like to share with the class about my invention are:

1. _____

2. _____

3. _____

What might you change about your magnet invention?

_____

_____

_____

MAGNETIC POLES

LEARNING INTENTION:

Students will identify where magnets are used in the environment and why they are used. Students will recognize that magnets have polarity, demonstrate that poles may either repel or attract each other and state a rule for when poles will repel or attract.

SUCCESS CRITERIA:

- sort through a variety of objects that use magnets and discuss why they are used.

- recognize where and how magnets are used in the environment.

- follow a procedure which demonstrates the effects of polarity on different kinds of magnets and communicate a rule for when magnetic poles attract or repel each other.

MATERIALS:

- copy of *Where Magnets Are Used* Worksheet for each student

- bar magnets, horseshoe magnets, doughnut-shaped magnets

- examples of magnets used in the house – an advertising or refrigerator magnet, a compass, a debit / ID card, shower curtain with magnets in bottom, electronic gear (such as sound speakers, open if possible to show magnet or elements inside)

PROCEDURE:

1. Begin a discussion with the class about where magnets are used. Explain that items with iron or steel in them are attracted to magnets. When people discovered this, they used magnets to help sort through things and separate out the items with certain metals. Magnets are used in all sorts of everyday items.

2. Go through the examples of magnets one by one. For example, ask students if they have ever seen someone use a debit / ID card. Draw their attention to the magnet strip on the back and discuss the role of using the magnet to store

security codes and information. Discuss other examples, such as industrial workplace magnets.

3. Give students *Where Magnets Are Used* Worksheet. Have them complete the worksheet. Take up the answers.

4. Bring the class together for a demonstration. Hold up two bar magnets. Highlight how each magnet is labelled with an 'N', north pole, and an 'S', south pole. Each magnet has a north pole and a south pole. Explain that if you take two magnets and try to push the North pole of one against the South pole of the other, you will feel a force trying to pull them together. We say, "Opposite poles attract each other."

5. Explain that if you take two magnets and try to push their south poles together, you will feel a force trying to push them apart. The same thing happens if you try to push the two north poles together. We say, "Like poles repel each other."

6. Demonstrate this attraction and repulsion with horseshoe magnets and with doughnut magnets as well.

7. Give students the *Where's the North Pole?* Worksheet, *North and South* Worksheet 1 and North and South Worksheet 2. Distribute 2 bar magnets to each group as needed. Read through the instructions with the class. Divide class into work groups or pairs as needed to complete the procedures. Monitor the progress of the class. Interview students and lead them to communicate the rule they find for when poles attract or repel each other.

8. Bring the class together to discuss doughnut magnets (or ring magnets). Demonstrate how doughnut magnets have the same North and South poles as other magnets, but on the top and bottom faces of the magnet. Explain to students they will label the poles on doughnut magnets (if needed) and then do an experiment with them. Distribute 2 doughnut magnets to each group along with the *Doughnut Magnet Poles* Worksheet 1 and *Floating Doughnut Magnet* Worksheet 2.

OTM-2171   ISBN: 978-1-4877-0199-4 © On The Mark Press

9. Bring class together in order to talk about what they found with the Floating Doughnuts. Let them use their imagination and ask how they could use floating doughnuts. Could they make toys, inventions, tools, or something else?

DIFFERENTIATION:

Slower learners can complete the Where Magnets Are Used Worksheet by cutting out and pasting the words in the correct spaces. For the magnetic poles exercises, pair up students to distribute work tasks.

For enrichment students can colour the items on the worksheet, or create a collage of items that use magnets from internet or magazine sources. This can be used as a room display or something for their own portfolios. As an extension, students could also draw designs of the brainstorming ideas from Step 9's final discussion.

OTM-2171   ISBN: 978-1-4877-0199-4 © On The Mark Press

Name: _____

Where Magnets Are Used

The items below use magnets. Can you name each one?

Words to Use

Microwave	Home magnet	Electric toothbrush
Construction crane	Computer	Compass
Sound speaker	Credit card	

_____

_____

_____

_____

_____

_____

_____

_____

OTM-2171 ISBN: 978-1-4877-0199-4 © On The Mark Press

Magnetic Poles

All magnets have a north pole and a south pole.

Bar magnet:
One end of the magnet is the north pole, the other the south pole.

Horseshoe magnet:
The poles are situated at either end.

Doughnut magnet:
The poles are situated on either face of the disk.

What Is Important about Poles?

If you take two magnets and try to push the north pole of one against the south pole of the other, you will feel a force trying to pull them together. We say that "opposite poles attract each other".

If you take two magnets and try to push their south poles together, you will feel a force trying to push them apart. The same thing will happen if you try to push two north poles together. We say that "like poles repel each other".

Sometimes, the poles are marked on bar and horseshoe magnets, but not always. North and South (pages 81-82) and Doughnut Magnet Poles (page 83-84) explain how to find the poles of unmarked magnets using a marked magnet.

OTM-2171   ISBN: 978-1-4877-0199-4 © On The Mark Press

Name: _____

Where's the North Pole?

Magnets have two ends. They are called the north pole and the south pole.

Draw a picture of your bar magnet. Write "**north pole**" and "**south pole**" at the right place.

OTM-2171   ISBN: 978-1-4877-0199-4 © On The Mark Press

North and South

You'll need:

- two magnets with poles identified

What to do:

1. Place the north pole of one magnet next to the south pole of the other magnet. Record what happens on your worksheet.

2. Place the north pole of the two magnets together. Record what happens on your worksheet.

3. Place the south pole of the two magnets together. Record what happens on your worksheet.

OTM-2171   ISBN: 978-1-4877-0199-4 © On The Mark Press

Name: _____

North and South

Let's Observe

When I put the north pole and south pole together:

_____

When I put the two north poles together:

_____

When I put the two south poles together:

_____

Let's Conclude

True or False?

When the two poles are the same, they push each other away.

Circle your answer: **True**   **False**

OTM-2171   ISBN: 978-1-4877-0199-4 © On The Mark Press

Doughnut Magnets

A doughnut magnet is shaped like the letter "O". The poles of the doughnut magnet are the top and bottom faces.

Cut out and glue the labels in the right place.

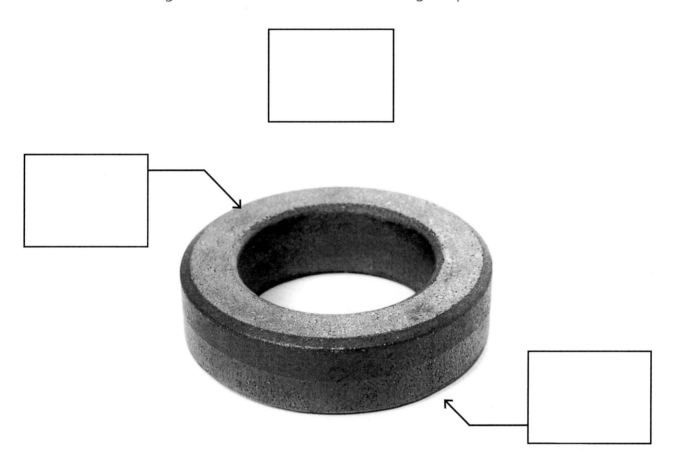

✂

| north pole | doughnut magnet | south pole |

Doughnut Magnet Poles

You'll need:

- a bar magnet with poles labeled
- doughnut magnet(s), at least one per team
- two white sticky labels per doughnut magnet

What to do:

1. Write "**north**" on one label. Write "**south**" on the other label.

2. The north pole of the bar magnet will attract the south pole of the doughnut magnet. Find the side of the doughnut that sticks to the north pole. Stick the **south** label to that side.

3. Label the other side with the **north** label.

4. Check that the **north** side of the doughnut is attracted by the south pole of the bar magnet.

OTM-2171   ISBN: 978-1-4877-0199-4 © On The Mark Press

Floating Doughnut Magnets

You'll need:

- two doughnut magnets with poles labelled
- a pencil or dowel

What to do:

1. Take a doughnut magnet, **north** side up.

2. Slide it around the pencil with the point facing down.

3. Take the other magnet, **south** side down.

4. Slide the other doughnut magnet around the pencil.

5. What happens? Draw a picture.

6. Flip over the top magnet.

7. What happens? Draw a picture.

Name: _____

Floating Doughnut Magnets

Let's Observe

1. Draw a picture of the first experiment, north side up:

[ ]

2. Draw a picture of the second experiment, north side down:

[ ]

Let's Conclude

3. What happens when you put two different poles together?

_____

_____

OTM-2171   ISBN: 978-1-4877-0199-4 © On The Mark Press

MAGNETIC FIELDS

LEARNING INTENTION:

Students will demonstrate an understanding of a magnetic field, recognizing that many materials are transparent to the effects of a magnet while others will interact with a magnetic field.

SUCCESS CRITERIA:

- sort through a variety of objects and determine which interact with the magnetic field and which do not
- follow procedures to test magnetic strength and accurately portray magnetic field patterns in drawings and in discussions

MATERIALS:

- copy of *Magnetic Strength* Worksheet 1 and *Magnetic Strength* Worksheet 2 for each student
- copy of *Uncover the Magnetic Field* Worksheet 1 and *Uncover the Magnetic Field* Worksheet 2 for each student
- copy of *What Changes the Magnetic Field* Worksheet for each student
- bar magnets, horseshoe magnets, doughnut-shaped magnets
- paper clips
- wooden rulers
- sticky tape
- iron filings
- aluminum foil
- magnetic field viewers (tray or stand)
- piece of white paper the size of the magnetic field viewer (if needed, to place on magnetic field viewer to make it easier to see the magnetic field)
- pencils, markers, pencil crayons

PROCEDURE:

1. Demonstrate to the class how the force produced by a magnet reaches some distance from the magnet. Hold a paperclip 1 cm away from a strong magnet and let go of it. The magnet still attracts the paperclip even if they don't touch. The paperclip still "jumps" towards it. But what happens if you hold the paperclip 10 cm away? Take predictions from the class and test them.

2. Tell students today they will be testing the magnetic strength of magnets and magnetic fields. Give out *Magnetic Strength* Worksheet 1 and *Magnetic Strength* Worksheet 2. Read through the instructions with the class. Divide up the class into pairs or work groups and give each group a bar magnet, sticky tape, paperclips and a ruler. Monitor their progress. Take up their work and discuss as a class as needed.

3. Bring the class together to discuss the idea of a magnetic field. The pattern in which the force reaches through the space around it is called a magnetic field. This pattern can be small or large. Set up the magnetic field viewer with a bar magnet underneath. (For demonstration purposes, a video such as www.youtube.com/watch?v=j8XNHlV6Qxg may be beneficial to the class).

4. Ask questions such as "What pattern do you see?" or "The iron shavings line up in what kind of shape?" or "What is happening around the poles?"

5. Tell students they will be investigators, uncovering the hidden magnetic field and testing to see what different objects affect it. Give students *Uncover the Magnetic Field* Worksheet 1 and *Uncover the Magnetic Field* Worksheet 2. Read through the instructions with them. Divide the class into groups or pairs and distribute the magnetic field viewers, iron filings and other test items. Monitor their progress with the worksheets.

6. Give students the *What Changes the Magnetic Field* Worksheet. Read through the instructions. Tell students they need to draw the results of each test.

7. Bring class together and lead a discussion on what conclusions can we make about magnetic fields. Do certain metal objects (paper clips or aluminum) change the pattern of the magnetic field? Do wooden objects (ruler) change the pattern? Do plastic items (marker) change the pattern? Do other magnets change the pattern?

DIFFERENTIATION:

For accommodation, some students can verbally communicate answers for writing tasks or worksheet questions.

Fast learners can develop artwork based on their discoveries of magnetic field patterns. As a further extension, students can research the Earth's magnetic poles and find out how much iron makes up the Earth's core.

For extension and vocabulary practise, students can complete the *Magnetic Wordsearch* worksheet included in this lesson.

OTM-2171   ISBN: 978-1-4877-0199-4 © On The Mark Press

Name: _____

Magnetic Strength

You'll need:

- a bar magnet
- sticky tape
- a paperclip
- a ruler

What to do:

1. How far do you think a magnet acts? Record your answer.

2. Tape the ruler to the middle of your desk.

3. Tape the magnet to the desk so that one of its poles touches the 0 cm mark of the ruler.

4. Hold a paperclip at the 5 cm mark. Let it go. What happens? Record your findings on the worksheet.

5. Repeat for 4 cm, 3 cm, 2 cm and 1 cm. Record your findings.

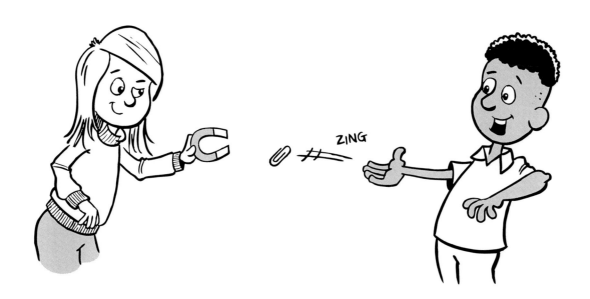

Name: _____

Magnetic Strength

Let's Predict

1. How far do you think a magnet acts?

_____

Let's Observe

2. Does the magnet attract the paperclip? Place an "X" in the correct column:

Distance	Yes	No
5 cm		
4 cm		
3 cm		
2 cm		
1 cm		

Let's Conclude

3. What is the maximum distance at which the magnet attracts the paperclip?

_____

OTM-2171   ISBN: 978-1-4877-0199-4 © On The Mark Press

Name: _____

Uncover the Magnetic Field

You'll need:

- a horseshoe magnet
- a bar magnet
- a magnetic field viewer toy

What to do:

1. Place the horseshoe magnet under the viewer.

2. What do you see? Draw a picture.

3. Place the bar magnet under the viewer.

4. What do you see? Draw a picture.

5. Compare your two pictures. What is similar? What is different? Uncover the Magnetic Field

Let's Observe

1. I placed the horseshoe magnet under the viewer. This is what I saw:

2. I placed the bar magnet under the viewer.
   This is what I saw: Uncover the Magnetic Field

Let's Conclude

3. What do the poles look like in the pictures?

_____

4. Can you see the shape of the magnet in the picture?

_____

5. What is different between the two pictures?

_____

_____

6. What is similar in the two pictures?

_____

_____

OTM-2171 ISBN: 978-1-4877-0199-4 © On The Mark Press

What Changes the Magnetic Field

1. I placed the bar magnet and a pile of paper clips under the view at the same time. This is what I saw. ⟶

2. I placed the bar magnet and a wooden ruler under the viewer at the same time. This is what I saw. ⟶

3. I placed the bar magnet and a ball of aluminum foil under the viewer at the same time. This is what I saw. ⟶

4. I placed the bar magnet and a marker under the viewer at the same time. This is what I saw. ⟶

5. I placed the bar magnet and the horseshoe magnet under the viewer at the same time. This is what I saw. ⟶

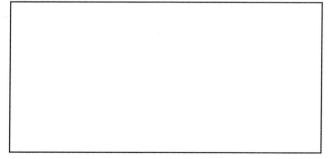

OTM-2171 ISBN: 978-1-4877-0199-4 © On The Mark Press

Name: _____

Magnetic Word Search

```
h  t  u  o  s  s  c  i  e
o  e  n  t  h  t  r  o  n
r  g  l  u  e  l  o  p  o
s  l  r  m  a  g  n  e  t
e  a  d  e  n  o  o  p  s
s  e  o  m  e  t  a  l  e
h  r  o  i  s  k  o  t  d
o  e  w  i  r  o  n  o  o
e  c  i  t  s  a  l  p  l
```

cereal	magnet	south	iron
lodestone	spoon	horseshoe	plastic
pole	Greek	north	wood
glue	metal	tool	

Leftover letters:

Would you like to be a _____?

OTM-2171   ISBN: 978-1-4877-0199-4 © On The Mark Press

TOPIC D: HOT AND COLD TEMPERATURES: MEASURING THE TEMPERATURE

LEARNING INTENTION:

Students will use a thermometer to measure temperature and learn to describe temperature using expressions such as hotter than or colder than.

SUCCESS CRITERIA:

- measure and compare temperature readings
- identify the clothing materials used to insulate people from the cold and to keep cool in the heat
- make predictions and gather findings about temperature, using charts and written descriptions
- make conclusions about temperature and make connections to people and the environment

MATERIALS NEEDED:

- a large thermometer (for instruction purposes)
- cocoa mix
- kettle
- thermometers, foam cups, spoons (3 of each per group)
- a jug of ice water, a jug of room temperature water, a jug of hot water
- a copy of **Hot and Cold Temperature** and **What to Wear?** Worksheets 1 and 2 for each student
- a copy of **A Cup of Cocoa** Worksheet 3 and 4 for each student
- pencils, markers

PROCEDURE:

* **Some direct instruction on how to read a thermometer may be necessary if the students have not yet acquired prior knowledge about temperature reading.**

1. Give students Worksheet 1 and 2 and allow them time to complete it on their own.

2. Students will investigate which temperature is the best for a cup of cocoa. Before beginning the experiment, it would be important to identify safe practises for handling hot materials and cold materials, and for avoiding potential dangers from heat sources such as boiling water.

*Handling or pouring of the hot water into the foam cups, should be teacher directed. Give students **A Cup of Cocoa** Worksheet 3 and 4, and the materials to do the experiment. Read through the question, materials needed, and what to do sections with the students. They can work in small groups to conduct the experiment, then make observations and conclusions.

DIFFERENTIATION:

Slower learners may benefit by working together in a small group with teacher direction to complete Worksheet 1 to ensure their comprehension of temperature reading.

For enrichment, faster learners could make cups of cocoa for another class in the school!

Hot And Cold Temperature

Thermometers measure the hotness and coldness of things.

Record the temperature shown on each thermometer.

Thermometer 1

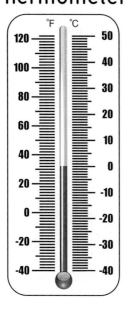

Thermometer 2

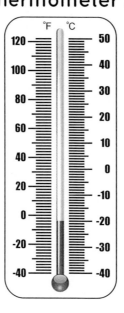

Thermometer 3

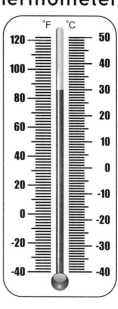

_____ °C   _____ °C   _____ °C

_____ °F   _____ °F   _____ °F

Use "**hotter than**" or "**colder than**" to compare the temperatures on the thermometers.

Thermometer 3 is _____ Thermometer 1.

Thermometer 3 is _____ Thermometer 2.

Which thermometer above could be used to describe the temperature in each picture?

Thermometer _____

Thermometer _____

OTM-2171   ISBN: 978-1-4877-0199-4 © On The Mark Press

Name: _____

What to Wear

Draw a line from each piece of clothing to its matching season.

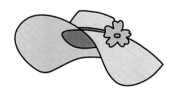

OTM-2171   ISBN: 978-1-4877-0199-4 © On The Mark Press

Name: _____

A Cup Of Cocoa

Question: Which temperature is the best for a cup of cocoa?

You'll need:

- cocoa mix
- a spoon
- 3 thermometers
- 3 foam cups
- ice water
- room temperature water
- hot water

What to do:

1. Make a prediction about the answer to the question. Record it on Worksheet 4.

2. Label the cups: "**Hot**", "**Medium**", and "**Cold**."

3. Pour hot water into the **Hot** cup, room temperature water into the **Medium** cup, and ice water into the **Cold** cup.

4. Place a thermometer in each cup and measure the temperature of the water in each cup. Record it on Worksheet 4.

5. Use the spoon to mix the cocoa and the water in each cup.

6. Write down how well the cocoa and water mixed together in each cup.

7. Taste each cup and write down how good it tastes.

8. Make conclusions about what you have observed.

OTM-2171   ISBN: 978-1-4877-0199-4 © On The Mark Press

Name: _____

Let's Predict

Which temperature is the best for a cup of cocoa?

_____

_____

Let's Observe

	Temperature in degrees	Did it mix well?	How did it taste?
Hot cup			
Medium cup			
Cold cup			

Let's Conclude

Which temperature of water **mixed the best** with the cocoa?

_____

Which temperature of cocoa **tasted the best**?

_____

In your opinion, which temperature makes the best cup of cocoa?

_____

INSULATORS

LEARNING INTENTION:

Students will learn how insulation is used to keep things hot or cold.

SUCCESS CRITERIA:

- identify materials that work as insulators to keep things hot or cold
- make predictions, gather findings about insulating properties of materials, using charts, drawings, and written descriptions
- make a plan to build an insulating device, make and record observations of the final product by comparing similarities, differences, and benefits of the design
- make conclusions about insulators and make connections to the environment

MATERIALS NEEDED:

- a foam cup, a paper cup, a plastic cup, a ceramic cup, a glass, 5 thermometers, a timer (a set for each group)
- kettle, and access to water
- a copy of *Insulators* Worksheet 1 and 2 for each student
- ice cubes, aluminum foil, a small plastic bag, an oven mitt (a set for each group)
- a copy of *Test an Insulator* Worksheet 3 and 4 for each student
- egg cartons, Styrofoam trays or plates, cardboard tubes, margarine tubs or other plastic containers, newspaper, cardboard, glue, string, tape, elastics, butterfly fasteners, paper clips, aluminum foil, popsicle sticks, pipe cleaners, scissors, hole puncher
- a copy of *Build an Insulating Device* Worksheet 5 and 6 for each student
- pencils, markers

PROCEDURE:

***This lesson can be done as one long lesson, or be divided into three shorter lessons.**

1. Students will investigate which materials make the best insulators by investigating which type of cup keeps water the hottest. Read through the question, materials needed, and what to do sections on Worksheet 1 with the students. They will make their predictions, then conduct the experiment in small groups. Using Worksheet 2, students will record results and make conclusions based on their observations.

2. Students will continue to investigate what material makes the best insulator. Give students *Test an Insulator* Worksheet 3 and 4, and the materials to do the experiment. Read through the question, materials needed, and what to do sections with the students. In groups of three, they will conduct the experiment, make observations and conclusions.

3. Explain to students that they are going to plan, design, and build an insulating device made out of some recyclable materials. They will compare their final product with a partner. Then, they will test their insulating device to see if it slows down the melting of an ice cube. Give each student a copy of *Build an Insulating Device* Worksheet 5 and 6 to complete, as they begin to build their invention. Once all insulators are completed, invite students to present their inventions to the class and share three things about it.

DIFFERENTIATION:

Slower learners may benefit by listing only one thing that is the same and one thing that is different from their partner's invention, and provide only one thing to share with the class about their own invention.

For enrichment, faster learners could build an additional insulator that works to keep things hot or cold that would be useful in their homes.

OTM-2171   ISBN: 978-1-4877-0199-4 © On The Mark Press

Name: _____

Insulators

Things that keep heat from transferring to the air are called insulators. Insulators also keep heat from getting inside. Insulators keep hot things hot, and keep cold things cold.

Question: Which kind of cup keeps water hot the best?

You'll need:

- a foam cup
- a paper cup
- a plastic cup
- a ceramic cup
- a glass
- 5 thermometers
- a stopwatch or timer

What to do:

1. Make a prediction about the answer to the question. Record it on Worksheet 2.

2. Your teacher will fill each of the cups with hot water.

3. Touch the outside of the cups. Write down how each feels to your hand on Worksheet 2.

4. Measure the temperature of the water in each of the cups. Record them on your worksheet.

5. Wait 15 minutes.

6. Measure the temperature of the water in each of the cups again. Record them on your worksheet.

7. Make conclusions about what you have observed.

Name: _____

Let's Predict!

Which kind of cup keeps water hot the best?

_____

Let's Observe

Type of Cup	How does it feel?	Temperature in degrees	Temperature after 15 mins.
Foam			
Paper			
Plastic			
Ceramic			
A glass			

Let's Conclude

Which kind of cup kept the water the hottest?

_____

Was this different than your guess? Explain.

_____

_____

Challenge: Which kind of cup would you use to keep your juice cold on a hot day? Explain your thinking.

_____

_____

OTM-2171   ISBN: 978-1-4877-0199-4 © On The Mark Press

Name: _____

Test An Insulator

Question: Which material is the best insulator?

You'll need:

- 3 ice cubes
- aluminum foil
- a small plastic bag
- an oven mitt
- a stopwatch or timer

What to do:

1. Make a prediction about the answer to the question. Record it on Worksheet 4.

2. Your teacher will divide your class into groups of three.

3. One person will wrap an ice cube in aluminum foil. The second person will put an ice cube into the plastic bag. The third person will put an ice cube into the oven mitt.

4. Wait 15 minutes.

5. Remove the ice cubes from the foil, plastic bag, and oven mitt.

6. Record your observations on Worksheet 4.

7. Make a conclusion about what you have observed.

OTM-2171   ISBN: 978-1-4877-0199-4 © On The Mark Press

Name: _____

Let's Predict!

Which material is the best insulator?

_____

Let's Observe

After 15 minutes, which ice cube was the largest?

_____

After 15 minutes, which ice cube was the smallest?

_____

Let's Conclude

Which material worked the best to slow down the melting?

_____

Were you surprised? Explain your thinking.

_____

_____

Draw a picture of a good insulator.

  OTM-2171   ISBN: 978-1-4877-0199-4 © On The Mark Press

Name: _____

Build An Insulating Device

Use what you know to design and construct a device to keep something hot or cold.

The materials I will use are:

_____   _____

_____   _____

_____   _____

My plan for making the insulating device is:

1. _____

2. _____

3. _____

4. _____

5. _____

Draw a design of what your insulator will look like:

Now gather your materials and begin your construction!

Name: _____

Share and Compare

Working with a partner, compare your final product.
List two things that are the same and two things that are
different about your insulators.

Things That Are The Same	Things That Are Different
1. _____ _____ _____ 2. _____ _____ _____	1. _____ _____ _____ 2. _____ _____ _____

Let's Test It!

Place an ice cube into your insulator. Place another ice
cube on a plate and put it on a table. Wait 15 minutes.
Record your observations by drawing a picture of what
happened to both ice cubes after 15 minutes.

Insulated ice cube	Ice cube on the table

Did your insulator slow down the melting of the ice cube?

_____

OTM-2171   ISBN: 978-1-4877-0199-4 © On The Mark Press

ADJUSTING THE TEMPERATURE

LEARNING INTENTION:

Students will identify ways in which the temperature in homes and buildings can be adjusted. Students will describe in general terms how local buildings are heated.

SUCCESS CRITERIA:

- identify a thermostat and decide to turn it up or down depending on descriptions of given situations
- decide to open or close a window based on descriptions of given situations and predicted results
- recognize how a space heater can adjust temperature in a cold room
- identify energy sources or fuel sources for heating buildings
- describe how heat is circulated through the school building and their own homes

MATERIALS NEEDED:

- a large thermometer (for instruction purposes)
- room thermostat or images of several kinds of thermostats
- kettle
- hair dryer
- open pipe with a turn in it (this can be galvanized or plastic, and doesn't need to be more than 30-50 cm in length total)
- a clamp or stand to hold the pipe above the kettle
- picture cards of energy sources – cut wood for wood burning, natural gas meter, a propane tank, a cylindrical oil tank, electrical outlet, solar panel, others as needed
- oven glove or heat protection to avoid touching hot kettle and pipe
- copy of *Adjusting the Temperature* Worksheet
- copy of *Home Energy Sources* Worksheet
- pencils, pencil crayons, markers

PROCEDURE:

1. Tell the students today you will discuss something found in almost every home and building. Ask them if they have ever heard the word "thermostat" before. Explain that a thermostat lets people adjust the temperature inside a building, either warming things up or cooling things down.

2. Show examples of thermostats. Ask students to raise their hands if they have one of the examples in their home (if a student doesn't know, that is fine).

3. Discuss other ways people adjust the temperature inside buildings – why do people open or close windows, use a space heater in a room, install an air conditioner in a window, etc.

4. Give students the *Adjusting the Temperature* Worksheet. Read the instructions with the class. Monitor student progress with the worksheet while they work independently. Students can colour the worksheet if they complete the work more quickly than others.

5. Prepare the class for a new discussion about energy sources. Tell students that we have many different ways to make heat for our homes. Ask students for examples of things in their homes that make heat. Some examples of leading questions might be, *"Do you have a stove in your house? Do you own a barbecue? Do you have a fireplace, a clothes dryer or a water heater?"* Explain that each appliance or machine produces heat. But how does each machine make heat? Does it have a plug for electricity? Does it use natural gas? Does someone put wood into it for burning?

6. Boil water in the kettle and ask students, *"Where does the steam from the kettle go?"* (It goes up). Ask, *"Since steam comes from the boiled water, is it very hot or very cold?"* (It is very hot). Safely check the temperature of the steam with a thermometer. Ask *"Does steam always try to go up?"* (yes) and *"What could we do to catch the steam?"* (accept reasonable answers)

8. Move the pipe over the kettle to demonstrate to the students how the steam can be redirected. Show how the steam goes into the pipe but then comes out in a new direction at the other end of the pipe. Carefully measure the temperature of the steam as it comes out the pipe with a thermometer for the class. Explain to the students that many heating systems heat up air or water and then move that air or water around using pipes. These pipes might go to every room in a house to heat up each room. It isn't just the air or water or steam that moves. The heat moves with it!

(Note: This demonstration can also be done with a hair dryer to show how heat and air circulate.)

9. Give students the **Home Energy Sources** Worksheet.

DIFFERENTIATION:

If learners need more practise reading thermometers, or if they complete their work properly ahead of others, they can try the thermometer game at **http://goo.gl/TSHZIM** . You may need to select the appropriate scale of thermometer for students before letting them work independently through the game.

As a homework extension, students can ask their parents how they heat their homes. Students will need to identify the energy source or fuel and report back to the class.

OTM-2171   ISBN: 978-1-4877-0199-4 © On The Mark Press

Name: _____

Adjusting the Temperature

Draw a line from each **Situation** to the matching **Action** you would take.

SITUATION

"It is very warm in here!"

"There is a nice breeze outside."

"It is very cold in here!"

"The house is warm but this one room is cold."

"It is too cold and windy inside the house."

ACTION

Turn up the thermostat.

Turn down the thermostat.

Open a window.

Close a window.

Turn on a heater.

OTM-2171   ISBN: 978-1-4877-0199-4 © On The Mark Press

Name: _____

Home Energy Sources

Draw a line from the machine or appliance that makes heat to the energy source it might use. Each appliance may use more than one energy source.

Stove

Furnace

Barbeque

Clothes Dryer

Water heater

Fireplace

Natural Gas   **Electricity**   **Oil**   **Wood Burning**   **Propane**

Can you think of another energy source?

_____

What do you use to heat your home?

_____

     OTM-2171   ISBN: 978-1-4877-0199-4 © On The Mark Press

HUMAN BODY TEMPERATURE

LEARNING INTENTION:

Students will record observations about human body temperature and understand a change in body temperature often signals a change in health.

SUCCESS CRITERIA:

- record observations and make predictions based on the evidence and information gathered.
- make conclusions about body temperature being relatively constant when someone is in good health

MATERIALS NEEDED:

- copy of *Body Temperature* Worksheet
- an ear thermometer or mouth thermometer
- sanitizing solution for thermometer
- pencils

PROCEDURE:

This procedure involves role-playing from the students and the teacher. Depending on the class and available resources, this setup can be adjusted so that students would work in groups instead.

1. Lead the class in a discussion about going to visit the doctor? Do you remember the last time you went to a doctor? Did the doctor take your temperature? Did the doctor ask you how you were feeling?

2. Explain that doctors need information in order to help solve problems. They often take a patient's temperature because human body temperature is relatively constant – it stays the same. A change in temperature can often signal a change in health. Something is different inside the body.

3. Present the class with the thermometer and tell them each student will get a chance to be a doctor. But they must first know how to take someone's temperature.

   To simplify the demonstration or process, the teacher can act as a "patient." Each student, one per day for a period of time, can then have a chance to act as "doctor." The doctor each day will look at the information on the chart (on the Body Temperature Worksheet), make predictions about today's temperature and then take the patient's temperature. The information will be recorded each day on the patient's chart. When possible, do this step at the same time each day for consistency.

4. After each student (or group) has had a chance to record information in the chart, review the observations with the class. Highlight how the body temperature remained relatively constant. Ask students about what they can predict about their own body temperature.

DIFFERENTIATION:

For slower learners, match students with partners so they can work as a team on the activities.

For enrichment, students can record the body temperature of a family member or themselves over a period of time in an independent project. They can use a copy of the chart from the worksheet or make their own. Then they can report to the teacher or class on what they discovered.

Name: _____

Body Temperature

Doctors measure the temperature of patients and ask patients questions to help them get better. They record what they observe on charts like the one on Worksheet 2.

What I Need

- thermometer
- pencil

What I Do

1. Look at the chart on the worksheet. Read the temperatures on each day. Make a prediction on what temperature the patient may have today.

2. Ask the patient if you can take their temperature. Use the thermometer as directed and measure the temperature from the thermometer. Record the temperature in the chart. Was your prediction correct?

3. Ask the patient, "How do you feel?" Write their answer in the chart.

4. Thank the patient.

5. According to your chart, answer these questions:

   Did the temperature go up and down a lot or stay almost the same?

   _____

   According to the observations, would you say the patient is in good health?

   _____

OTM-2171   ISBN: 978-1-4877-0199-4 © On The Mark Press

Name: _____

Let's Observe

Week	Monday	Tuesday	Wednesday	Thursday	Friday
1	Temperature: ____ °C How do you feel? ____	Temperature: ____ °C How do you feel? ____	Temperature: ____ °C How do you feel? ____	Temperature: ____ °C How do you feel? ____	Temperature: ____ °C How do you feel? ____
2	Temperature: ____ °C How do you feel? ____	Temperature: ____ °C How do you feel? ____	Temperature: ____ °C How do you feel? ____	Temperature: ____ °C How do you feel? ____	Temperature: ____ °C How do you feel? ____
3	Temperature: ____ °C How do you feel? ____	Temperature: ____ °C How do you feel? ____	Temperature: ____ °C How do you feel? ____	Temperature: ____ °C How do you feel? ____	Temperature: ____ °C How do you feel? ____
4	Temperature: ____ °C How do you feel? ____	Temperature: ____ °C How do you feel? ____	Temperature: ____ °C How do you feel? ____	Temperature: ____ °C How do you feel? ____	Temperature: ____ °C How do you feel? ____

TEMPERATURE CHANGES IN OUR DAILY LIVES

LEARNING INTENTION:

Students will identify materials that insulate animals from the cold as well as identify materials that are used by humans for the same purpose.

Students will describe ways in which temperature changes affect us in our daily lives.

SUCCESS CRITERIA:

- communicate an understanding that different materials help protect animals and human beings from cold temperatures.
- demonstrate how activities and clothing can change due to changes in temperature.

MATERIALS NEEDED:

- samples of different articles of clothing – parkas, jackets, shorts, bathing suits, t-shirts, winter toques, ear muffs, and more as needed.
- animal images or illustrations – bear, sheep, a furry dog, a snowy owl, a duck, whale, seal and more as needed
- pencils, markers, crayons
- copy of *Protection from the Cold* Worksheet for each student
- copy of *Hot and Cold in our Daily Lives* Worksheet for each student

PROCEDURE:

1. Explain to the class they will play a game called *Hot or Cold*. One student will come to the front of the class. The teach will secretly tell them to pretend that it is either hot or cold outside. That student then has to pick and wear one of the articles of clothing. Then when it is the class's turn to guess, the teacher will ask, "Is it cold outside?" and "Is it hot outside?" Students will raise their hand to vote and the teacher will keep track of cold and hot votes. Go through 2-5 rounds of volunteers for the game.

2. Tell the class we could do the same thing for animals. Only this time, instead of clothing, we need to identify what the animal has that protects it from the cold. Use one of the animal images

as an example. Ask, "What kind of animal is this?" "What kind of coat does it have to protect it from the cold?" Go through 2-5 rounds of animal images so that many different examples of coverings such as fur, feathers, wool and blubber are mentioned. Explain each word as needed. Blubber, for example is a layer of tissue inside the body that provides both insulation and energy.

3. Give students the *Protection from the Cold* Worksheet. Read the instructions with the students. Read the list of words they can use to complete the sentences. Give students time to complete the worksheet. Monitor student work.

4. Start a discussion with the class about daily temperatures. Do you wear the same thing outside in winter as you do in the summer? Why do you change what you wear? What are some examples of things you would wear when it is cold? What activities do you do when it is cold outside? What do you wear in the summer? What activities do you do when it is warm outside? Explain that it is temperature changes that affect what we choose to do and what we wear.

5. Give students the *Hot and Cold in our Daily Lives* Worksheet. Read the instructions with the students. Give students time to complete the worksheet.

DIFFERENTIATION:

Slower learners can be paired with other students and worksheets can be completed together as needed. Other accommodations may include finishing the worksheets verbally instead of in written form.

Faster learners can research Arctic animals, migrating animals or shedding animals to understand how their coats and daily activities are affected by temperature changes through the seasons.

OTM-2171   ISBN: 978-1-4877-0199-4 © On The Mark Press

Name: _____

Protection from the Cold

Finish each sentence by writing the correct word on the line. Draw a picture for each sentence.

jacket   feathers   socks   wool   fur   blubber

A sheep grows a coat of _____ to stay warm.

A duck grows a coat of _____ to keep warm.

A bear grows a coat of _____ to keep warm.

I put on a _____ to stay warm.

A whale has a layer of _____ to stay warm.

I put on a thick pair of _____ to keep warm.

OTM-2171   ISBN: 978-1-4877-0199-4 © On The Mark Press

Name: _____

Hot and Cold in our Daily Lives

Finish each sentence by listing the things you would wear or the activities you like to do. Draw a picture for each sentence.

When it is cold outside, I wear

_____

_____

_____

When it is cold outside, I like to

_____

_____

_____

When it is warm outside, I wear

_____

_____

_____

When it is warm outside, I like to

_____

_____

_____

OTM-2171   ISBN: 978-1-4877-0199-4 © On The Mark Press

TOPIC E: SMALL CRAWLING AND FLYING ANIMALS: INVERTEBRATES

LEARNING INTENTION:

Students will recognize that there are many different kinds of small crawling and flying animals, by identifying a range of examples of invertebrates. Students will describe conditions for the care of small animals and demonstrate responsible care in maintaining the animal for a period of time.

SUCCESS CRITERIA:

- identify the body parts of an insect, a spider, a worm and a crustacean
- list types of common insects
- create a home for earthworms
- record observations about earthworm activity
- describe the growth and changes in a hermit crab over time

MATERIALS NEEDED:

- a copy of *What's an Invertebrate? Part One* Worksheet 1, 2, 3 and 4 for each student
- a copy of *What's an Invertebrate? Part Two* Worksheet 5 and 6 for each student
- a copy of *What's an Invertebrate? Part Three* Worksheet 7 for each student
- a copy of *A Home for Earthworms* Worksheet 8 and 9 for each student
- a copy of *What's an Invertebrate? Part Four* Worksheet 10 for each student
- a copy of *A Home for a Hermit Crab* Worksheet 11 and 12 for each student
- access to the internet
- soil such as sand and loam or topsoil, enough to fill large jars for each student
- a hammer and a nail, masking tape, a jug of water, a few small cups
- earthworms (2 to 3 per student)
- vegetable or fruit scraps
- large sheets of black construction paper (2 per student)

- chart paper, markers, pencil crayons, clipboards, pencils
- optional materials - painting paper, paint, brushes, modelling clay
- optional materials – visit a local pet store or go on line to gather the materials and information needed to create a home for a hermit crab

PROCEDURE:

*This lesson can be done as one long lesson, or be divided into shorter lessons.

1. Using Worksheets 1 and 2, do a shared reading activity with the students. This will allow for reading practise and breaking down word parts to read the larger words. Along with the content, discussion of vocabulary words would be of benefit for their comprehension.

   Vocabulary words: invertebrate, head, antennae, facets, complex eyes, exoskeleton, thorax, organs, image, simple eyes, insect, abdomen, mandibles

2. Give students worksheet 3 and 4. They will list 5 insects that they know, and then compare lists with a classmate. Students may need to access the internet or visit a local library to find out more about the physical appearance of the insect they choose to draw and label, and to find out what it looked like upon hatching.

3. Using worksheets 5 and 6, do a shared reading activity with the students. This will allow for the reading practise and breaking down word parts to read the larger words. Along with the content, discussion of vocabulary words would be of benefit for their comprehension. Students will also need to access the internet or visit a library to find facts about spiders. An option is to come back as a large group and have students share their facts.

   Vocabulary words: arachnids, fangs, poisonous glands, silk, vibrations, cephalothorax, abdomen, spinnerets, venom, prey

4. Using Worksheets 5 and 6, do a shared reading activity with the students. This will allow for the reading practise and breaking down work parts to read the larger words. Along with the content, discussion of vocabulary words would be of benefit for their comprehension. Students will also need to access the internet or visit a library to find facts about spiders. An option is to ocme back as a large group to have students share their facts.

Vocabulary words: slither, burrows, moist, digests, earthworm, soil, vegetation, castings, creature, crumbly, tunnels, nutrients

5. Students will create a home for earthworms. Give them Worksheets 8 and 9 along with the materials needed. Read through the sections on the worksheets – materials needed, what to do – to ensure their understanding of the task. Students will make and record observations of the earthworms home as it is created and 48 hours later. They will make a conclusion about the purpose earthworms have in creating nutrient-rich soil.

6. Using Worksheet 10, do a shared reading activity with the students. This will allow for reading practise and breaking down word parts to read the large words. Along with the content, discussion of vocabulary words would be of benefit for their comprehension.

Vocabulary words: crustaceans, head, abdomen, environment, exoskeleton, jointed, thorax, antennae, dangers, protects

7. Together as a class, create a home for a hermit crab that can be maintained in the classroom. You will need to visit a pet store or go online to gather the information on the materials you will need and how to care for it daily.

8. Once the hermit crab's home is created, give students Worksheet 11. They will illustrate what it looks like.

9. After a few weeks or months, give students Worksheet 12. They will detail how they cared for the hermit crab and comment on any changes they have noticed.

DIFFERENTIATION:

Slower learners may benefit by working in a small group with teacher support to orally answer the questions on Worksheet 12. Responses could be recorded on one large chart paper, and then displayed in the classroom or on a bulletin board. This chart paper could then be surrounded by an art activity (painting) done by students that depicts the hermit crab in its environment.

For enrichment, faster learners could use modelling clay to create three dimensional models of the four types of invertebrates (insect, spider, worm, crustacean). These could be left on a display table or used in an art project.

OTM-2171   ISBN: 978-1-4877-0199-4 © On The Mark Press

Name: _____

What is an Invertebrate? Part One

An invertebrate is an animal with no backbone. Some invertebrates like insects have an exoskeleton, which is a skeleton that is on the outside of their bodies.

Insects have bodies that are made up of three parts. These parts are the head, the thorax and the abdomen. They have three pairs of legs that grow out from the front, middle, and back of the thorax. The head of an insect has two antennae.

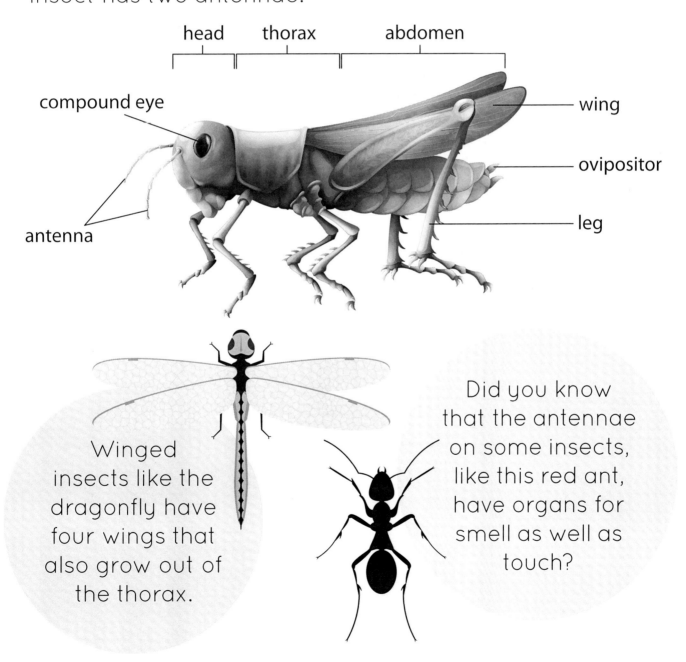

head   thorax   abdomen

compound eye

wing

oviposter

leg

antenna

Winged insects like the dragonfly have four wings that also grow out of the thorax.

Did you know that the antennae on some insects, like this red ant, have organs for smell as well as touch?

Name: _____

Did you know that insects can have a pair of complex eyes and simple eyes too? Let's learn more about this!

Complex eyes have many small lenses called facets, and they can see colour. The parts that each facet sees

comes together to make one sharp image. So, insects see very well with their complex eyes.

A housefly has about 4000 lenses inside of each of its compound eyes.

If an insect has simple eyes too, they will be found in a form of three on top of its head. These eyes are simple because they only see light and dark, not colour.

A form of 3 simple eyes sit on the top of this paper wasp's head.

Fast Fact!

Many insects have mandibles, which are a pair of jaws. The jaws of an insect close sideways, unlike our own which close up and down.

OTM-2171   ISBN: 978-1-4877-0199-4 © On The Mark Press

1. List 6 insects that you know

_____   _____

_____   _____

_____   _____

2. Compare your list with another classmate's list. Name
   2 insects from their list that are not on yours. (Or think
   of 2 more insects you know.)

_____   _____

Did You Know?

All invertebrates lay eggs. That means all the insects on
your list are egg laying creatures!

For some insects, when they hatch out of the egg
they look like their adults. For example, when an ant is
hatched, it looks like a small ant.

For some insects, when they hatch out of the egg, they
go through changes before they look like their adults. For
example, when a caterpillar is hatched, it will become a
pupa before it turns into an adult butterfly.

OTM-2171   ISBN: 978-1-4877-0199-4 © On The Mark Press

Choose one insect from your list.

1. Draw it in the box below.

2. Label its antennae, head, thorax, abdomen, legs, eyes and mandibles.

3. Label its wings if it has some.

Draw what your insect looked like when it hatched from it egg.

OTM-2171   ISBN: 978-1-4877-0199-4 © On The Mark Press

Name: _____

What is an Invertebrate? Part Two

Another group of invertebrates are arachnids. Arachnids are creatures that have bodies made up of two parts. They have eight legs. They do not have wings or antennae. Can you guess what common creature is part of this group? It is the spider!

A spider, like all other invertebrates, does not have a backbone. It has an exoskeleton. As the spider grows, it sheds its first exoskeleton and then stretches out as a new, larger exoskeleton forms.

The front part of a spider is called the cephalothorax.

The cephalothorax has the spider's eyes, brain, mouth, fangs and stomach. A spider's eight legs grow out of this part of its body.

If a spider is poisonous, this is where its poisonous glands would be.

The back part of the spider is called the abdomen. A spider's spinnerets are at the back end of the abdomen. A spider makes silk in its spinnerets. Some spiders use this silk to spin webs.

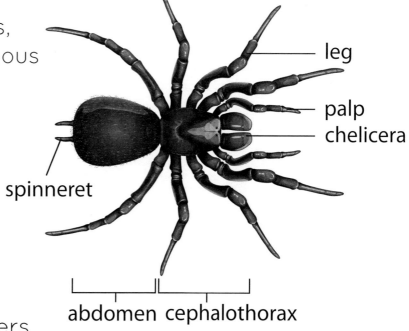

Did you know that a spider's body has oil on it that help it from sticking to its own web?

Name: _____

Spinning Out The Facts!

Did you know that a spider has 48 knees? Each one of its legs has 6 joints on it! At the end of each of its legs, a spider has small claws.

There are many hairs that cover a spider's legs. These hairs can sense vibrations or movements. They can also pick up smells in the air.

Some spiders have venom in their fangs. They use the venom in their fangs to catch their prey. The shot of venom stills their prey, and then the spider goes in for a tasty snack!

Some spiders have poisonous venom. If a spider's prey gets poked by its fangs, the poison will turn the prey's guts into liquid. Then, the spider will suck it out!

Now it is your turn to spin out the facts! Use the internet or visit a local library to learn about spiders. Find and tell two facts about spiders.

1. _____

_____

_____

_____

2. _____

_____

_____

_____

     OTM-2171   ISBN: 978-1-4877-0199-4 © On The Mark Press

What is an Invertebrate? Part Three

Worms are also invertebrates. They have no legs, so they slither or inch their bodies along to move from place to place. There are different kinds of worms, but the one that is most common is the earthworm. Let's learn more about this creature!

Where Do They Live?

Earthworms make their homes in soil. Some earthworms live in the roots of grass and some live in vegetable or flower gardens. Some earthworms will even make their burrows under leaves around tree roots.

The earthworm's burrow is long and dark. Earthworms like to stay in their burrows because it is dark and moist.

The earthworm comes out at night to eat leaves, grass or other vegetation.

What Do They Do?

Some invertebrates are plant eaters. They eat living plants. Some eat other animals. The earthworm is called a decomposer. The earthworm decomposes or recycles material from dead plants and animals The earthworm eats and then digests its food. Its castings add nutrients to the soil that new plants need to grow.

"Worms help water flow through the soil!"

A Home for Earthworms

Some people make worm farms to make rich soil for their gardens at home. Let's give this a try!

You'll Need

- a large jar with a lid
- a hammer and a nail
- masking tape
- 2 or 3 earthworms

- 2 sheets of black construction paper
- soil (loam and sand)
- vegetable or fruit scraps
- a small cup of water

What To Do

1. Fill the jar three-quarters full of loam soil. Add a layer of sand on top.

2. Put some vegetable or fruit scraps on top of the sand.

3. Moisten the soil with a bit of water. Then add the earthworms.

4. Using the hammer and a nail, **your teacher** will make some holes in the lid of the jar

5. Put the lid on the jar to close it.

6. On worksheet 4, draw what you see in the jar.

7. Cover the sides of the jar with black construction paper so that no light can get in. Earthworms like it dark!

8. Leave the jar for 2 days.

9. After 2 days, take off the paper. Record what you see now on Worksheet 9.

OTM-2171   ISBN: 978-1-4877-0199-4 © On The Mark Press

Name: _____

Let's Observe

Draw your observations of what is in the jar.

This is what it looked like in the jar **before it was covered up**:

This is what it looked like in the jar **after it was left for 2 days**:

Let's Conclude

What did the earthworms do?

_____

_____

_____

_____

Name: _____

What is an Invertebrate? Part Four

Another group of invertebrates are crustaceans. They live mostly in the water. Some examples of crustaceans are lobsters, crabs and shrimp.

Crustaceans have many legs with jointed parts. Some crustaceans like shrimp and crabs can swim, but others like lobsters must use their legs to move from place to place.

The body of a crustacean is made up of the head, the thorax, and the abdomen. Its eyes are in the middle of its head. Did you know that crustaceans have two pairs of antennae? They use their antennae to see what is in their environment, like food, or dangers.

Did You Know?

A crustacean's exoskeleton protects its body. Its body will grow, but its exoskeleton will not. So, it will shed its shell when its body gets too big for it, and grow another one. Except for the hermit crab! The hermit crab cannot make his own shell, so it hides in shells left behind by other crustaceans.

OTM-2171   ISBN: 978-1-4877-0199-4 © On The Mark Press

Name: _____

A Home for a Hermit Crab

Work together with your teacher and classmates to set up a home for a hermit crab. Watch what happens as the hermit crab grows. Record what you see.

This is the home that we made for the hermit crab.

OTM-2171   ISBN: 978-1-4877-0199-4 © On The Mark Press

Name: _____

Let's Observe!

What do you do to care for the hermit crab so that its needs are met?

1. _____

_____

_____

_____

2. _____

_____

_____

_____

3. _____

_____

_____

_____

What growth and changes have you noticed in the hermit crab?

_____

_____

_____

_____

OTM-2171 ISBN: 978-1-4877-0199-4 © On The Mark Press

A STUDY OF INVERTEBRATES

LEARNING INTENTION:

Students will compare and contrast small animals that are found in the environment. Students will recognize that small animals have homes where they meet their basic needs of air, food, water, shelter and space, as well as describe special characteristics that help the animals survive in its home. Students will identify an animal's role in the food chain. Students will identify and give examples of ways that small animals avoid predators.

SUCCESS CRITERIA:

- choose an invertebrate to research
- describe its appearance, where it lives, what it eats, its predators, its day and night time activities, and special changes it experiences
- compare and contrast two invertebrates
- display all information using pictures, diagrams and written descriptions
- orally share one interesting fact about an invertebrate

MATERIALS NEEDED:

- copy of *Investigating an Invertebrate* Worksheet 1, 2, 3 and 4 for each student
- copy of *Investigating an Invertebrate* Worksheet 5 and 6 for each student
- access to the internet or local library
- Bristol board (½ piece per student), or display board
- chart paper, markers, glue sticks, pencils, clipboards, pencil crayons
- light coloured construction paper, scissors, (optional materials for trivia card creation)

PROCEDURES

*This lesson can be done as one long lesson, or done in two or three shorter lessons.

1. Engage students in a brainstorming activity to name different types of invertebrates that they know. Record student ideas on chart paper. Ask students about what each invertebrate brings to the food chain. With each example, ask what animal might eat the invertebrate (frogs may eat several kinds of invertebrates). Tell students invertebrates may use several methods to avoid predators – speed, flying, burrowing or even camouflage. Lead students to find examples for each.

2. Instruct students to choose an invertebrate that they would like to learn more about. Give students Worksheets 1, 2, 3 and 4 with a clipboard and pencil. With access to the internet or resources at a local library, students will investigate and record information about their chosen invertebrate. Upon completion, give each student a half piece of large Bristol board or display board so they can present their work. Be sure to review vocabulary from the last lesson and cover new vocabulary with the class.

   Vocabulary Words: plant eater, meat eater, decomposer, burrow, migrate, molt, camouflage, predator

3. Divide students into pairs (ensuring their chosen invertebrates are different species). Give students Worksheet 5 and 6 with a clipboard and pencil. They will compare and contrast their invertebrates. This will allow for further learning about other species of invertebrates.

4. Invite students to participate in a "Did you know?" activity. They will come back to the large group, with one interesting fact that they would like to share with the rest of the class. Instruct them to form their sentence about their interesting fact as "Did you know...?" This activity will promote rich discussion and use of life science terminology about invertebrates.

5. Display Bristol boards containing student work around the school or in the classroom.

DIFFERENTIATION:

Slower learners may benefit by working as a small group with teacher support and direction to complete Worksheets 1, 2, 3 and 4. One invertebrate should be chosen, and then to lessen the work load, each student in the group could be assigned a section to research, and then share the findings with the small group in order to complete the worksheets.

For enrichment, faster learners could create a trivia card game using the facts that they learned from the "Did you know...?" activity. Trivia questions should begin with "What invertebrate...?"

OTM-2171   ISBN: 978-1-4877-0199-4 © On The Mark Press

Name: _____

Investigating an Invertebrate

You have learned about different types of invertebrates. Now it is time to choose one to learn more about what it looks like, where it lives, what it eats and how it lives. Let's get started!

The invertebrate I am investigating is

_____

This is a diagram of what it looks like.

Where does it live?

_____

_____

_____

_____

Name: _____

Make a list of things that your invertebrate eats.

_____ _____

_____ _____

_____ _____

_____ _____

How does it get its food?

_____

_____

_____

My invertebrate is a (circle the best answer)

a plant eater   **an animal eater**   **a decomposer**

Draw and label some of its predators.

     OTM-2171   ISBN: 978-1-4877-0199-4 © On The Mark Press

Name: _____

Some animals avoid their predators by taking cover in burrows, by running away with great speed, by flying away, or by using camouflage. How does your invertebrate avoid predators?

My invertebrate _____

Complete the sentences and add drawings.

In the **daytime**, my invertebrate

_____

_____

_____

_____

At **night time**, my invertebrate

_____

_____

_____

_____

Name: _____

Tell about any special changes that happen to your invertebrate. For example:

- **Does it migrate?**

- **Does it change its form?**

- **Does it molt?**

_____

_____

_____

_____

_____

Wrapping It Up!

By investigating this invertebrate, I learned...

_____

_____

_____

_____

I still wonder about...

_____

_____

_____

_____

OTM-2171   ISBN: 978-1-4877-0199-4 © On The Mark Press

Name: _____

The Invertebrate Exchange!

Partner up with a classmate. Exchange information about the invertebrates that you have investigated. Be sure that your classmate's invertebrate is different from yours!

This is the invertebrate that my partner investigated.

The _____

Do your invertebrates look the same in any way? Explain.

_____

_____

_____

_____

Do your invertebrates look different in any way? Explain.

_____

_____

_____

_____

This is what is on the menu for my partner's invertebrate:

_____

_____

_____

_____

_____

Circle any food items that are the same for your invertebrate.

My partner's invertebrate is a... (circle the best answer)

a plant eater   **an animal eater**   **a decomposer**

Talk with your partner about the day and night time activities of your invertebrates.

Our invertebrates both like to _____

_____

_____

My invertebrate likes to:

_____

_____

_____

but my partner's invertebrate likes to:

_____

_____

_____

OTM-2171   ISBN: 978-1-4877-0199-4 © On The Mark Press

FITTING THE ENVIRONMENT: ADAPTATIONS OF SMALL ANIMALS

LEARNING INTENTIONS

Students will identify ways in which animals are considered helpful or harmful to humans and to the environment. Students will recognize some animals must adapt to extreme conditions to meet their basic needs. Students will describe the relationship of these animals to other living and non-living things in their habitat, and recognize that small animals have homes where they meet their basic needs of air, food, water, shelter and space.

SUCCESS CRITERIA:

- illustrate the migration in a Monarch Butterfly's life in relation to avoiding extreme conditions.

- read and discuss how the arctic woolly bear moth makes a home to meet its basic needs and how it adapts to extreme conditions in its environment.

- discuss the mosquito's relationship to other living and non-living things in its habitat and whether the frog is helpful or harmful to humans and to the environment.

MATERIALS NEEDED:

- a copy of *Monarch Butterfly Migration* Worksheets 1 and 2 for each student

- a copy of *Arctic Woolly Bear Caterpillar* Worksheets 3 and 4 for each student

- a copy of *The Mosquito* Worksheets 6 and 7 for each student

- access to the internet and local library

- sheets of art paper (optional)

- pencils, pencil crayons, markers, glue, scissors, rulers

PROCEDURE:

*This lesson can be done as one long lesson, or done in shorter lessons.

1. Do a read aloud about butterflies. A suggested read aloud is *From Caterpillar to Butterfly* by Deborah Heiligman. Alternatively, use a video to introduce the class to monarch butterfly migration. A suggested video is *https://youtu.be/LawHWsIqa5s*. Afterward, review some of the points highlighted in the reading or video to focus on migration. Students could engage in a "turn and talk" activity. With a partner, they can talk about what piece of information from the story or video they found most interesting. Give students Worksheets 1 and 2 to complete and monitor their work.

2. Tell the class that some animals have to do some amazing things to survive. Many beetles and flies hibernate, or go into a very deep sleep in the winter. But the Arctic Woolly Bear Caterpillar must survive after being completely frozen! As an option, show the class the video "Caterpillar Survives Frozen Death" *http://goo.gl/5vXoHE*. Give students Worksheets 3 and 4. Read through the material with the class and monitor the students as they complete the work.

3. Lead the class in a discussion about mosquitoes. Who has ever been bitten by one? What do they look like? Are they helpful to human beings or are they hurtful? What role does a mosquito have if they are hurtful? Review the parts of an invertebrate (thorax, head, abdomen, etc) in relation to a mosquito. Continue the discussion with a talk of the mosquito life cycle (egg, birth, growth, adulthood) and how if we asked those same questions to frogs and birds we might get a different answer. As an option, play the following video, or something similar, for the class "Mosquito life Cycle" *https://youtu.be/wFfO7f8Vr9c*. Give out Worksheets 5 and 6. Read through the worksheets with the class and give students time to complete the work. Monitor their progress.

DIFFERENTIATION:

Slower learners may benefit by working as a small group with teacher support and direction to complete worksheets. Allow students to complete questions on the worksheets with recorded verbal responses.

For enrichment, faster learners could research how long it takes a monarch butterfly to complete its migration, as an extension of worksheets 1 and 2. Students could find and report on another arctic insect after worksheets 3 and 4. Students could create a Mosquito Preparation Kit after Worksheets 5 and 6, and present an advertisement for the kit, featuring the benefits of the kit, the solutions it provides or why people should buy their kit.

OTM-2171   ISBN: 978-1-4877-0199-4 © On The Mark Press

Name: _____

Monarch Butterfly Migration

Some animals **migrate**. This means they travel long distances in order to live somewhere else for a while. Some animals do this because they want to avoid cold weather. Another reason might be to have their babies in a safer place. Animals might even migrate so they can find food.

Monarch butterflies migrate every winter. The wings of the monarch butterfly are often less than 10 cm long. But the little insects travel more than 4000 kilometres!

Use a ruler and draw a line near the top of the box that is 10 cm long.

Under that line, Draw a picture of a monarch butterfly.

Name: _____

Below is a map. The darker dots show places monarch butterflies live in the summer. The white dots show the places where monarch butterflies migrate for the winter. Draw a line to connect the dark dot in the north to the white dot in the south.

How far do the monarch butterflies travel when they migrate ?

_____ kilometres

OTM-2171   ISBN: 978-1-4877-0199-4 © On The Mark Press

Arctic Woolly Bear Caterpillar

The Arctic Woolly Bear Caterpillar lives in Northern Canada. As a caterpillar, it will feed off the leaves of any plants it can find. The caterpillar must survive the cold in the Arctic, where sometimes winter can last eight months of the year. Plants in the arctic can be very hard to find and very small.

The Arctic Woolly Bear Caterpillar hides between rocks in the winter. The entire body of the caterpillar goes very still and freezes. But when the ice and snow melt, the caterpillar is still alive! Even though it can get very cold, the caterpillar has adapted so that nothing inside or outside is damaged by freezing.

During the summer, they bask in the sunshine a lot of the time in order to warm up. Most of their life is spent in hibernation or lying very still.

The caterpillar has **predators**. Some wasps and flies land on the caterpillars to lay their eggs on or inside the caterpillars. The caterpillar grows long "hairs" to help protect it from predators, but the long hairs also help it keep heat inside its body.

It often takes seven years or more for the caterpillar to eat enough and grow strong enough to finally build a cocoon and turn into an adult moth.

OTM-2171   ISBN: 978-1-4877-0199-4 © On The Mark Press

Name: _____

How long is winter in the Arctic? _____ months.

How many years will it take for an arctic woolly bear caterpillar before it turns into a moth?_____ years.

Draw a picture of the arctic woolly bear caterpillar and a picture of the arctic woolly bear moth.

Did You Know?

Other insects hibernate in the winter as well. For example, beetles look for loose bark to burrow under for hibernation. They also spend the winter under piles of fallen leaves for warmth and safety.

Wrapping It Up!

By investigating the arctic woolly bear caterpillar, I learned...

_____

_____

I still wonder about...

_____

_____

OTM-2171   ISBN: 978-1-4877-0199-4 © On The Mark Press

Name: _____

Mosquitoes!

Mosquitoes fly in the air but where are they born? Mosquitoes lay their eggs in water. The life cycle of a mosquito has four stages:

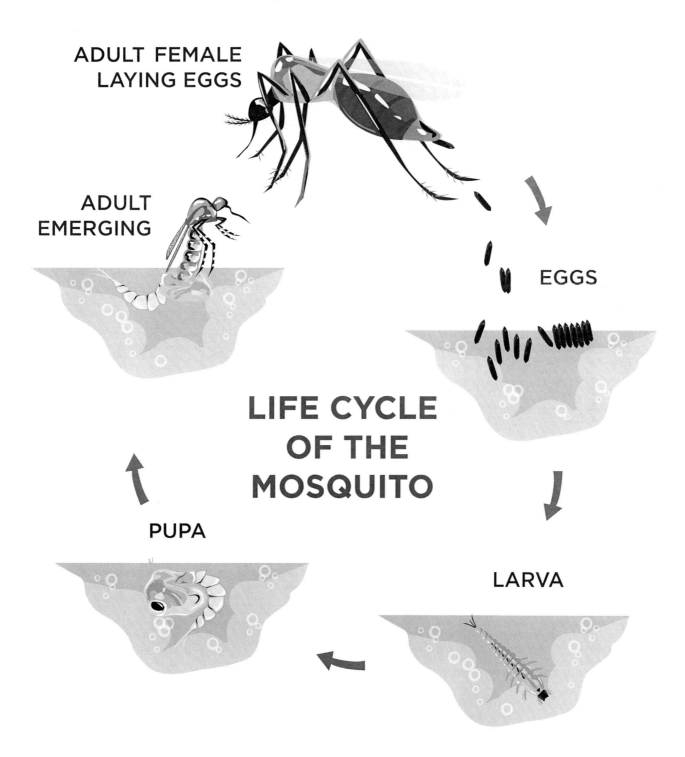

ADULT FEMALE LAYING EGGS

ADULT EMERGING

EGGS

LIFE CYCLE OF THE MOSQUITO

PUPA

LARVA

Mosquitoes create a lot of problems for human beings. In some parts of the world, diseases are spread through mosquito bites. Here is a list of some diseases that mosquitoes have spread:

Malaria	common in tropical parts of the world
Dog Heartworm	harmful to dogs and cats
West Nile Virus	harmful to birds, rare in humans but can lead to infection with other diseases
Zika Virus	rare disease that can cause birth defects and nerve damage

Mosquitoes have many predators everywhere. Birds such as ducks and swallows eat adult mosquitoes. Other insects like dragonflies eat mosquito larva. Fish eat mosquitoes in every stage of their life cycle.

Even though mosquitoes might be harmful to human beings, many animals depend on them for their food.

Mosquitoes need water in order to survive and grow. People can stay safe from mosquito bites if they wear

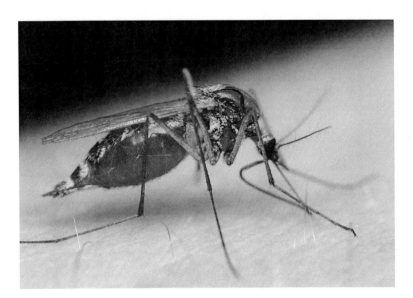

clothes that cover the skin, use a repellent and be careful around mosquito habitats like ponds, marshes and places with open water.

OTM-2171   ISBN: 978-1-4877-0199-4 © On The Mark Press

Here is a diagram of an adult mosquito. Label the following words on the diagram.

Abdomen   thorax   antennae   leg   eye   head

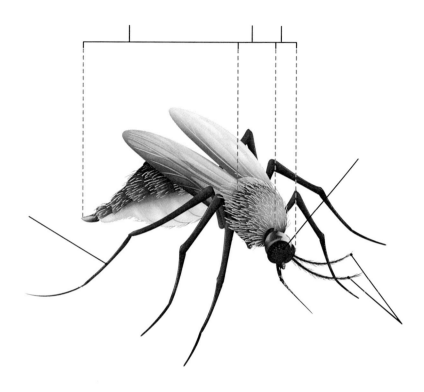

Wrapping It Up!

What do mosquitoes need in order to survive?

_____

Name 3 predators of mosquitoes.

_____   _____   _____

Name 3 things people can do to protect themselves from mosquito bites.

_____

_____